Newnes

Electronics Assembly Pocket Book

Keith Brindley

NEWNES

Newnes
An imprint of Butterworth-Heinemann Ltd
Linacre House, Jordan Hill, Oxford OX2 8DP

PART OF REED INTERNATIONAL BOOKS

OXFORD LONDON BOSTON
MUNICH NEW DELHI SINGAPORE SYDNEY
TOKYO TORONTO WELLINGTON

First published 1991

British Library Cataloguing in Publication Data
Brindley, Keith
  Newnes electronics assembly pocket book
  A catalogue record for this book is available
  from the British Library
  ISBN 0 7506 0222 8
Produced by Sylvester Publications, Loughborough

Printed and bound in Great Britain by Courier International, East Kilbride

Preface

This book is a result of collaboration between the Engineering Training Authority (ETA) — formerly known as the Engineering Industry Training Board (EITB) — Butterworth-Heinemann the publishers, Sylvester Publications the typesetting and design agency, and myself as editor. In total, it represents considerable work over many years; researching, compiling, writing, editing and publishing work relevant to all aspects of electronics assembly. Indeed, in this form, it is merely the latest variation of earlier presentations.

Newnes Pocketbooks are intended as concise guides to designated subject areas. Newnes Electronics Assembly Pocket Book is by far the most concise such work covering electronics assembly ever published. Yet it is by no means too simplistic. Topics are covered to a sufficient depth to give readers a basic, though thorough grounding. Further, there are just a handful short of one thousand illustrations here.

Newnes Electronics Assembly Pocket Book was produced and typeset electronically in 8 pt Times using Apple Macintosh computers. Original text was input using Word version 4, while pages were laid out with Aldus Pagemaker version 4. Camera-ready copy was presented equivalent to 450 dpi.

Contents

Safety and safe working practice

Interference
Unauthorised interference with electrical apparatus is extremely dangerous. Do not take liberties with electricity. Never interfere with levers, knobs, push-buttons or switches on machinery or electrical equipment which you have not been trained to operate.

Authorised working
Before working on electrically powered equipment:

● switch off the power supply and make sure it cannot be switched on accidentally
● make use of cautionary notices
● disconnect from supply if possible
● earth high-voltage circuits or capacitors to remove static charges
● if equipment is live, beware of non-standard colour-codes used for wiring
● check wiring or components with a voltmeter or neon test lamp before touching equipment.

When using electrically powered hand tools, do not leave the supply cable trailing; this might lead to accidents.

Working conditions
Following practices and conditions in the work situation make a vital contribution to health and safety, while adding to efficiency:

● adherence to place-of-work safety regulations
● good quality lighting
● efficient ventilation
● air conditioning
● controllable temperature
● clean room conditions
● wearing eye protection (safety goggles)
● wearing correct protective clothing which has been well-laundered
● use of safety guards
● use of correctly serviced machinery.

Working practices
In addition to place-of-work regulations, following good working practices should be observed:

● faulty electrical equipment must not be used and must be reported immediately
● electrical equipment must not be touched with wet hands
● frayed wires are dangerous and must be reported
● power tools must never be connected to a lamp socket as these have no earth connection. Even when a power tool is double insulated it should not be connected to a lamp socket as power tools demand higher currents than lighting circuits are designed to carry; this is a fire risk
● become conversant with use of firefighting equipment and its location

● ensure a second person is present when working with electrical power

● learn positions of all emergency exits and emergency power-off switches

● memorise telephone number or extension number of supervisor and emergency services. In the event of accident, give assistance and call for help

● ensure the case of any power unit being used is correctly earthed

● ensure the ventilation system is working correctly, especially when soldering or using adhesives and solvent

● avoid inhaling fumes from solder, adhesives, solvent and so on

● confirm all power has been switched off before soldering on a chassis

● stow a hot soldering iron on its stand and ensure the stand is in a safe position on the workbench

● do not allow a soldering iron lead to trail over the workpiece. Remove all other obstructions which might restrict movement of the lead

● never stow a hot soldering iron in a toolbox

● during continuity testing, switch off power

● always take time; avoid haste

● ensure the workbench is clean and tools are arranged in an orderly manner

● arrange components and parts in a logical sequence, using protective containers where specified

● stow sharp instruments and tools in a safe place

● do not use defective tools. Avoid using the wrong tool

● follow handling instructions for small or touch-sensitive components

● do not operate unfamiliar equipment without first reading its instruction manual

● ensure safety guards are in working order

● never use a part or component which is not specified

● clear all extraneous equipment from workshop; chairs, tool carriers, equipment stowage trays and so on.

Emergencies or accidents
Personal injury
Call for immediate help. Treat bleeding and shock.

Fire
Know position of nearest exit, alarm, and extinguisher. Note: special extinguishers are required for electrical fires.

Electric shock
Switch off power, or drag victim away from live parts using non-conducting material such as a length of wood, or dry material such as a scarf. Administer artificial respiration and send for expert help.

Safety in electrical work
Danger from electricity is covered by No 28 of the Electricity (Factories Act) Special Regulations 1908 (Competent Persons Exemption) Order 1968. Work on equipment with exposed conductors live at voltages greater than 250 Vdc or 125 Vac may only be undertaken by an authorised person, or by a competent person under the former's supervision, provided that such a

competent person is either over the age of 21 or over the age of 18 and in at least the third year of an appropriate course. This means that a trainee not over 18, although in the third year, cannot perform such work — however well supervised.

Safety in electrical work must be a predominant theme at all times, rather than being looked on as an isolated activity. Repeated reference should be made to IEE Regulations, Statutory requirements and Codes of Practice. Codes of Practice may differ between various factories. Before commencement of any work which entails use of mains supplies, a knowledge of following items is essential:

● existing factory/work rules and codes of practice
● comparison between domestic and industrial conditions and requirements
● voltages available at socket outlets. Use of distinguishing sockets and plugs
● voltage for supply of portable equipment
● reasons for earthing
● which equipment is earthed
● reasons for double insulation
● which equipment is double insulated
● recommended action to be taken in event of electric shock.

Safe use of electricity
Note the colour-code for 3-core cables used with power tools and so on:

live	brown
neutral	blue
earth	green/yellow

Ensure these are connected correctly.

If an electrical cable moves freely at its point of entry to a plug or appliance, or if the individual insulated wires are visible, a dangerous condition is developing. See it is corrected immediately. Check continuity of earth wires at regular and frequent intervals. Report all defects at once, for example:

● cracked or perished insulation
● loose pins on plugs
● badly fitted plugs and sockets
● detached earth wires.

If defects are not removed they tend to be accepted as normal. They become ignored. They cause accidents.

Electrical fire risks
Observe all regulations relating to safe use of electricity and to electrical fire risks. Fires may be started by:

● overheating of cables, appliances and plugs
● sparks produced under operating conditions, say, by contactors or commutators
● sparks which result from a circuit being broken.

In the case of connections and terminals there is an electrical fire risk if:

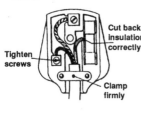

Tighten screws

Cut back insulation correctly

Clamp firmly

● cable current-carrying capacity is insufficient
● capacity of the plug is inadequate
● insulation is cut back too far
● a conductor is damaged while insulation is cut back
● connections are not tight
● cable is not adequately supported at the point of entry to a plug or appliance.

Shroud

When a reinforcing rubber shroud is provided, ensure it is used.

Always fit a suitably rated bulb to an inspection lamp. Use of a higher-powered bulb may overheat the lampholder and damage insulation. Do not rest the lamp on flammable materials as the high surface temperature of a bulb may cause a fire.

First aid

Procedure in event of accident

Important

Minor injuries, however trivial, should be dealt with in a first aid room. This is particularly important for removal of foreign bodies from the eye and for treatment of burns caused by hot metal or slag. Ensure you are familiar with the injuries reporting procedure as laid down by your employer. Ensure you understand your responsibilities in respect of both your own safety and that of all other persons in the vicinity of your working area, especially as defined in the Health and Safety at Work Act, 1974.

General

No unqualified person should attempt to render first aid except to try to save a life in extreme emergency when:

● breathing has stopped
● bleeding is severe.

Everyone must know the procedure for obtaining medical aid, or for summoning a doctor and ambulance. At all times the position of the nearest telephone and the nearest first aid box must be known.

First aid to save life
If breathing has stopped, apply immediate artificial respiration.

Mouth-to-mouth artificial respiration

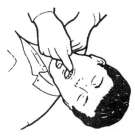

1   Remove any obstructions from the mouth.

2   Support the nape of the neck and press the top of the head backwards.

3   Press the angle of the jaw forward from behind to ensure the tongue is forward and the airway clear.

4   Open your mouth wide and take a deep breath. Pinch the victim's nostrils. Seal your lips around his mouth. Blow gently until his lungs are filled. Remove your mouth and watch the victim's chest movement. When exhalation is complete, repeat procedure until patient breathes by himself.

If an assistant is available, heart massage may be employed to assist recovery by improving circulation of blood.

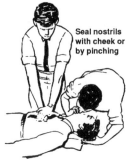

Seal nostrils with cheek or by pinching

Support nape of neck

1   Kneel alongside the victim's chest.
2   Place the heel of the right hand on the lower end of the victim's chest bone.
3   Place the left hand on top of the right hand.
4   Keep your arms stiff and rock forwards so your weight depresses the victim's chest by 3 to 5 cm (not more than 5 cm).
5   Repeat this movement steadily at one-second intervals. Erratic or violent action is dangerous.

When working alone, heart massage may still be applied while carrying out mouth-to-mouth resuscitation.

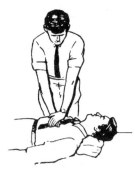

1   Kneel alongside the victim's chest.
2   Carry out inflation a few times as previously described.
3   Change to the heart massage position and depress the chest fives times at one-second intervals.
4   Give another inflation.
5   Repeat heart massage five times. Continue to alternate with one inflation and five chest depressions until help arrives.

Treatment of electric shock

Prompt treatment is essential. Avoid direct contact with the victim until electrical power is turned off. If assistance is close at hand, send for medical aid then carry on with emergency treatment. If you are alone, proceed with treatment at once.

Switch off current if this can be done without undue delay. Otherwise, using dry non-conducting material such as a wooden bar, rope, a scarf, the victim's coat-tails, any dry article of clothing, a belt, rolled-up newspaper, non-metallic hose, PVC tubing, remove victim from contact with the live conductor.

Electrical burns

A person receiving an electric shock may also sustain burns when the current passes through his body. Do not waste time by applying first aid to the burns until breathing is restored and the patient can breathe normally and unaided.

Burns and scalds

Burns are very painful. If a large area of the body is burnt, give no treatment except to exclude air eg, by covering with water, clean paper, or a clean shirt. This relieves the pain. If a burn has been caused chemically by an acid or an alkali, the chemical must be washed off with clean water.

Wash off chemicals with water

Severe bleeding

Any wound which is bleeding profusely, especially in the wrist, hand or fingers, must be considered serious and must receive professional attention. As an immediate first aid measure, pressure on the wound itself is the best means of stopping the bleeding and avoiding infection.

Immediate action

Always in cases of severe bleeding:

Raise injured part

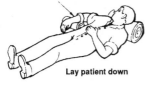

Lay patient down

1　Make the patient lie down and rest.
2　If possible, raise the injured part above the level of the body.
3　Apply pressure to the wound.
4　Summon assistance.

To control severe bleeding

Stop bleeding with pressure at the wound

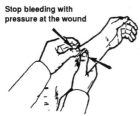

Squeeze together the sides of the wound. Apply pressure as long as it is necessary to stop the bleeding. When the bleeding has stopped, put a dressing over the wound, and cover it with a pad of soft material.

　　For an abdominal stab wound, such as that caused by falling on a sharp tool, keep the patient bending over the wound to stop internal bleeding.

Large wound

Apply a clean pad (preferably an individual dressing)and bandage firmly in place. If bleeding is very severe apply more than one dressing.

Fire procedures

General

When fire is discovered, immediate correct action is essential to provide the best possible chance of putting it out quickly, thus reducing danger to life and damage caused.

All personnel must be fire conscious and must know the fire drill, position of the nearest fire point and the nearest telephone and alarm.

Whenever equipment which could give rise to a fire hazard is used, such as blowlamps or welding torches, a suitable fire extinguisher should be placed within easy reach.

Fire drills and precautions

Fire drills and orders should be displayed on all section notice boards and at all fire points. All personnel must make themselves familiar with those orders, which should cover the following points:

- sounding the alarm
- informing the Fire Brigade
- evacuation of premises
- assembly
- roll-call
- fire-fighting, pending arrival of the Fire Brigade.

Fire exits

Fire exits and escape routes should be clearly marked and must be kept free from obstructions. You should always know the way to the nearest fire exit.

Raising the alarm

Raise the alarm even for small fires

As soon as fire occurs, raise the alarm.

Immediate attack

Provided there is no personal danger involved, a fire should be attacked immediately on discovery with the first aid fire-fighting equipment available.

Safety
Stop fighting the fire and leave the area if:

● the fire appears to be beyond control, or
● the escape route is threatened by fire, or
● smoke obscures or threatens to obscure the escape route.

Allocation of fire-fighting equipment

Each department must hold at least the legal minimum amount of fire-fighting equipment, as specified on the fire certificate.

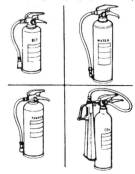

The equipment held must be of the correct type to deal with any type of fire which can be expected. Such equipment must be maintained in serviceable condition and placed so it is readily available for use.

All extinguishers in one premises should, where possible, operate by the same method. Where different types of extinguisher are sited together they must be properly labelled to prevent the wrong type being used.

Fire points

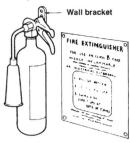

Wall bracket

Where possible fire extinguishers should be supported by brackets firmly fixed to the wall at a convenient height at all fire points. The following information should be clearly displayed at each fire point:

● type of fire for which the extinguisher is suitable
● how to operate the extinguisher.

These details should appear either on the appliance itself or on an instruction plate mounted close to the extinguisher.

Manual lifting techniques

Correct manual lifting techniques

How the spine straightens

Correct lifting techniques must always be used. The human spine is not an efficient weight-lifting machine and is easily damaged if incorrect techniques are used. Following technique should be used:

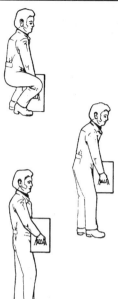

● the lift should start with the lifter in a balanced squatting position, legs slightly apart and the load to be lifted held close to the body. Ensure a safe secure hand grip is obtained. Before the weight is taken, the back should be straightened and held as near the vertical position as possible

● to raise the load, straighten the legs. This ensures the lifting strain is correctly transmitted and is taken by the powerful thigh muscles and bones

● to complete the lift, raise the upper part of the body to the vertical position.

Principal rules for holding and carrying

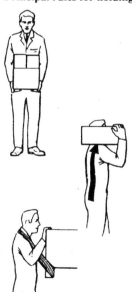

Following rules apply:

● keep the body upright when carrying a load and hold the load close to the body
● distribute the weight of the load evenly

● let the bone structure of the body support the load

● use aids such as harnesses, straps and yokes.

Protection against injury

Risk of injury and damage to equipment is reduced if:

Clear movable objects

● the area of operation is clear of obstructions
● all movable objects are moved to a safe place.

Standards and specifications

Various standards apply to work, governing the way equipment is made, performs and is tested. Standards can be issued by customers, the manufacturing company, national standards organisations, and international standards organisations.

Customer standards

These standards are usually called specifications and are issued by any of the following:

● Government departments; Ministry of Defence, UK Atomic Energy Authority and so on
● major consulting engineering organisations
● individual customers - British and overseas.

Specifications normally involve relevant national standards 'as far as they apply' and include specific product requirements relating to constructional details, fittings, dimensions, finish and so on. Product specified may be:

● a component used in an assembly
● the assembly itself, or a part.

A description of the process recommended to achieve product requirements may be included.

All specifications attempt to establish a degree of standardisation and, hence, interchangeability of products from different manufacturers, produced to equal quality and performance levels.

Company standards

These are established by individual manufacturers and are normally conveyed to the production facility by means of standard sheets, data sheets and so on. Such documents are concerned with tolerances, clearances, quality of finish, method of performing various operations and processes and so on.

Company standards are identified by various design authorities within a company:

● design department
● drawing office
● planning department.

National standards

In general, manufactured goods comply with relevant British Standards, issued by the British Standards Institution. They are mainly concerned with quality and performance, rather than details of manufacture and are widely used throughout industry. They describe standard ranges of products and so provide a product identification, readily understood by the:

● designer, who is saved the task of preparing specifications for materials chosen
● purchaser, who can order to a standard code
● supplier, who knows the standard of product which must be supplied.

A list of available standards and other relevant documents is given annually in the British Standards Institution catalogue. Occasionally, particularly where products are manufactured for sale overseas, relevant national standards for those overseas countries are used instead of British Standards.

International standards

Internationally, electrical standards are issued by the International Electrotechnical Commission (IEC). Such standards are in force world-wide. Often, however, British Standards are direct equivalents of international standards and following one implies following the other.

Use of standards and specifications

Read standards and specifications carefully. Pick out those sections which are relevant to the work in hand and ensure they are understood. Obtain and read additional standards and specifications where specified. If contradictions are apparent, seek advice. Ensure standards and specifications have the correct issue number, and check with the design authority if in doubt. Check all standards and specifications are up-to-date.

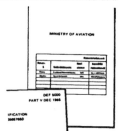

Note: many Government standards are marked with a date-of-issue rather than an issue number.

Example of standards use

To determine correct sizes of fixing holes for components on a printed circuit board:

● as the board is used as telecommunications equipment for the armed services, it is governed by DEF5000
● refer to DEF5000 and locate the section on printed wiring (Section IX: mechanical design, general construction, wiring and lubrication)

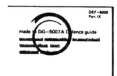

● the passage on printed wiring refers to specification DG5007A: defence guides for printed wiring of electronic telecommunications applications

● refer to defence guide DG5007A. Section 6 deals with fixing holes for components and provides all information required.

Engineering information

Engineering drawings

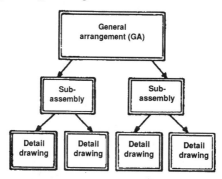

Many engineering drawings are produced for a particular equipment. The general arrangement drawing shows a complete view of the assembled equipment. Each sub-assembly in the equipment has a unique drawing showing a complete view of that sub-assembly. In electronics a sub-assembly is frequently called a module. Detail drawings show details of parts of the equipment.

A general arrangement drawing and sub-assembly drawing has the following drawings and documents associated with it:

● circuit diagram
● block diagram
● wiring diagram
● wiring schedule
● component layout
● sequence chart
● specifications and data sheets
● parts list.

Circuit diagram

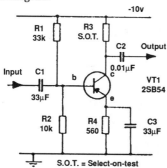

A circuit diagram shows, by symbols, components within the circuit and their interconnections. It does not show exact physical positions of parts. Reference should be made to relevant standards eg, British Standards:

● BS308 Engineering drawing practice
● BS3939 Graphical symbols for electrical power, telecommunications and electronics diagrams.

Block diagram

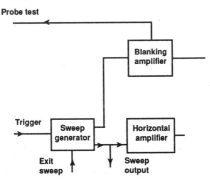

In the case of complex equipment block diagrams are provided which show units within the equipment in stage format. Such diagrams provide an understanding of the theory of operation without the detail essential for manufacture.

Wiring diagram

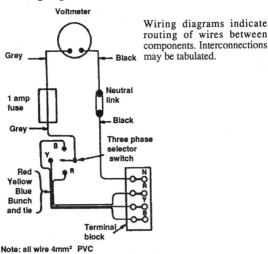

Wiring diagrams indicate routing of wires between components. Interconnections may be tabulated.

Wiring schedule

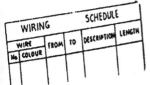

A wiring schedule (sometimes called a run-out sheet) lists wires used in numerical order of assembly, and provides all information for selecting wire and laying it in.

Wiring layout

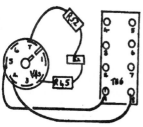

Wiring layouts show the physical layout of wiring and often include details of colour coding or numbering of wire ends.

Component layout

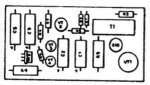

A component layout drawing indicates positions of components on a printed circuit board (PCB).

Sequence chart

SEQUENCE CHART		
OP. No.	PROCESS INSTRUCTIONS	PROCESS SPEC. No.
1	Cut and strip all P.V.C. wire reqd. for this assy.	47
2	Twist and tin all ends	47
3	Fit markers as reqd.	54

A sequence chart indicates the order in which operations must be performed and techniques used.

Specifications and data sheets

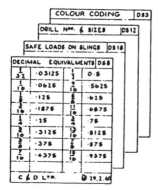

Specifications define what a product must be able to do (design parameters), how it is to be tested after production, and so on. Data sheets provide general information over a wide range.

Parts list

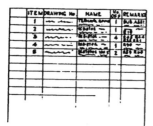

A parts list tabulates all parts and consumables (solder, wires, adhesives and so on) required in fabrication of an assembly and its sub-assemblies. This data forms the basis for illustrated parts lists (IPLs) issued by most manufacturers. Such 'exploded' illustrations are of great help when building or servicing equipment.

Drawings, documentation and practices

Drawing control

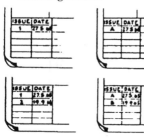

Each set of drawings for a particular product is allocated a unique block of numbers. Normally the general arrangement drawing is given the first number and others follow. All documents incorporate an issue number and date. When a change is made to a product, drawings affected are modified, issue numbers are changed and documents are re-issued at the new date.

Important
Unless otherwise specified, it is essential that all old issues of drawings are discarded and only latest issues are used.

Working from drawings or sketches
A number of procedures should be followed before starting work when using a drawing or dimensioned sketch:

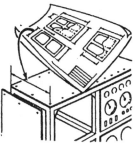

● check the projection and scale of the drawing
● confirm the drawing is reliable by checking dimensions or connections given against existing equipment
● check all units of measurement are stated correctly
● confirm all features are shown
● confirm all dimensions or instructions required are shown
● obtain answers to all queries before starting work.

While working to drawings, procedures to be followed are:

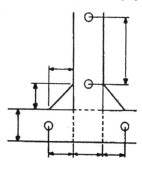

● continually cross-check that figures and dimensions are not misread
● work from datum points where possible, to avoid cumulative errors in measurement
● note that measurements scaled from drawings may be unreliable
● if errors are found, note them and refer back to the design authority.

Freehand sketching

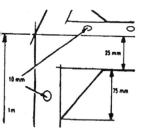

If parts are to be made or an installation planned it is advisable to draw a sketch before starting work. This is useful in planning the work and calculating materials needed. If details are shown correctly a freehand sketch need not be exactly in proportion. However, it must be clear and neat to prevent error in reading from it. Symbols used should follow relevant standards eg, BS308 and BS3939.

When making a sketch:

● use clean paper and a sharp pencil or ballpoint pen. Rest the work on a flat surface or support it firmly
● keep the drawing as clean and clear as possible

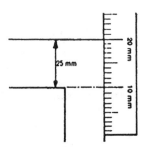

● make rough sketches to decide what views are needed
● when complete, check carefully against existing objects for omissions or errors
● take measurements where appropriate and mark on sketch

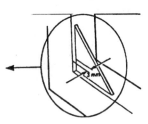

● show wiring connections with colour-coding and identification markers
● if complicated or very small, make an enlarged sketch of detail
● make final check before leaving the work location

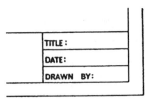

● if necessary, re-draw in good proportion especially if the drawing is to be used by others
● title, date and sign the drawing.

Authorised drawings should be available for all major work, or for larger assemblies and installations.

Preparing diagrams

Where a suitable diagram of equipment does not exist and is required for modification, repair, or fault-finding, it must be prepared by inspection of the actual equipment. A suggested procedure is:

● check all circuits are isolated from the mains supply or other source
● identify components of the circuit
● where necessary, identify connections between components by using appropriate test equipment, and mark these on the diagram
● mark on the diagram relevant circuit component and wiring identifications.

Definition of units

Voltage, current and resistance

Voltage

Voltage is that force which tends to cause a movement of electrons in a closed circuit. Unit of voltage is the volt (V), where one volt is the difference in potential between two points in a circuit carrying a current of one ampere when the power dissipated between the two points is equal to one watt.

Current

Electric current is a flow of electrons where the greater the rate of flow, the larger the current. Unit of current is the ampere (A), where one ampere is defined as being the current if a charge of one coulomb passes a given point each second.

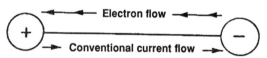

Direction of flow is shown.

Resistance

Resistance is the opposition to current flow. It is a property of materials which impedes the passage of electrons. Unit of resistance is the ohm (Ω), where one ohm is defined as being the resistance if an applied voltage of one volt causes a current of one ampere to flow.

Electromotive force and potential difference

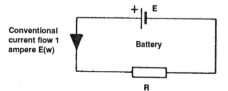

In a simple DC circuit an electromotive force (emf) of E volts causes a current of one ampere to flow against the resistance of R ohms. In a simple electrical circuit the battery voltage is referred to as emf. This corresponds to battery voltage before a resistor is connected. When the circuit is completed, a potential difference (pd) measured in volts exists across the ends of the resistor. Value of pd is less than that of emf due to battery internal resistance.

Ohm's law

This law states that current through a resistor varies in direct proportion to voltage applied across a resistor; provided temperature remains constant. The law is applied to purely resistive DC and AC circuits.

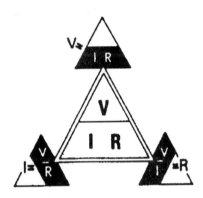

From this can be derived:

$$I = \frac{V}{R}; \quad V = IR; \quad R = \frac{V}{I};$$

where I is current in amperes, R is resistance in ohms, V is applied potential difference in volts.

Resistivity

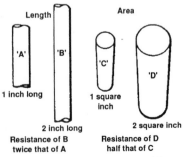

Resistance increases with length, decreases with cross-sectional area and depends on nature and composition of the material.

Temperature coefficient of resistance

Resistance of most materials alters with temperature change. Pure metals increase in resistance with a rise in temperature, the change for some alloys being quite small. For many insulators, on the other hand, resistance decreases considerably with an increase in temperature.

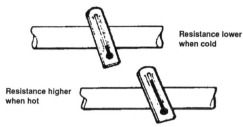

Resistance lower when cold

Resistance higher when hot

Temperature coefficient of resistance has a numerical value equal to change in resistance which one ohm of material undergoes when temperature is increased from 0°C to 1°C.

Power in direct current (DC) circuits

Power is the rate of doing work or transforming energy. Power in DC circuits is given by V x I = power in watts (W) or joules per second, where V is applied voltage in volts and I is current in amperes.

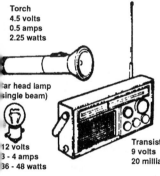

Torch
4.5 volts
0.5 amps
2.25 watts

Car head lamp
(single beam)

12 volts
3 - 4 amps
36 - 48 watts

Transistor radio
9 volts
20 milliamps

From Ohm's law V = IR, therefore power = VI = IR watts. This emphasises that when current through a resistor is doubled, power consumption increases fourfold.

For some equipment (eg, motors) power is quoted as horsepower (HP), where:

1 HP = 746 W.

BS3939 Electronics symbols

BS3939 is the national standard which recommends symbols, with definitions, for use in electrical power, telecommunications and electronics diagrams. It comprises graphical symbols for general application, symbol elements and qualifying symbols. It corresponds exactly with international standard IEC617. BS3939 is organised into a number of parts, each dealing with a particular subject or area.

Resistors

Graphical symbols – general

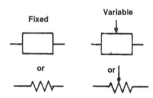

Fixed

Variable

or

or

The resistor is one of the most important components used in electronic circuits. Tasks of the resistor include controlling and directing current flow, making changing currents produce changing voltages (as in a voltage amplifier) and producing variable voltages from fixed ones (the potential divider).

Tolerance

Exact values cannot be guaranteed by mass production methods but this is not a problem as, in most electronic circuits, values of resistors are not critical. Tolerance tells us minimum and maximum values we can expect a resistor to have and is expressed as a percentage above and below the stated (or nominal) value eg, a resistor of nominal value 100Ω with a tolerance of ±10% has a value between 90Ω and 110Ω.

Stability

This is the ability of a component to keep the same value as it ages, despite changes of temperature and other physical conditions. This can be an important factor in some circuits.

Power rating

When resistors function they dissipate heat. Rate at which heat is dissipated determines maximum current they can withstand. This is given by the power rating. If a resistor exceeds its power rating it overheats and is damaged or destroyed. For most electronic circuits 0.25W or 0.5W power ratings are adequate. Generally, greater the physical size of a resistor, greater is its rating.

Resistor types

Carbon resistors

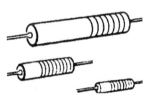

Carbon composition resistors are made from a mixture of carbon (a conductor) and clay (a non-conductor) which is pressed and moulded into rods by heating. Resistance per unit volume depends on the ratio of conducting to insulating material. Values range from a few ohms to 10MΩ with a typical tolerance of ±10% and ratings from 0.125W to 2W. They have poor stability but are cheap.

Carbon film resistors are made by depositing a thin film of carbon on a ceramic rod and etching a spiral track through the film. Values, ratings and cost are similar to carbon composition resistors but tolerances are better (±5%) and stability is very good.

Metal film resistors

These overcome the problems associated with carbon resistors, they are also capable of being manufactured to very close tolerances in the order of 0.1% as opposed to 1 to 2% for carbon although this, in itself, does not justify the relatively high cost of manufacture. They are used in critical applications such as navigational systems and medical instruments although these by no means form all their applications.

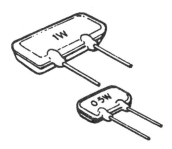

Construction of metal film resistors is similar to that of carbon film resistors. Power ratings are generally smaller than carbon types - typically 0.125W to 1W.

Both carbon and metal film types have resistance values painted on the body, usually conforming to a colour-code. Maximum power dissipation can be determined by physical size. Sizes of 0.125W metal film resistors are usually about 4mm long by 2mm in diameter, while 2W resistors are about 24mm long by 5mm.

Wirewound resistors

These are wound with wire having a very low temperature coefficient of resistance; typically nickel chromium. Wire is wound onto a ceramic former then coated with a ceramic glaze. They can be made to very close tolerances but their main advantage is high power rating, typically 2W to 20W. If the resistor is aluminium-clad, power rating can be as high as 50W.

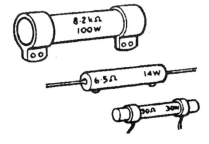

Resistance value is either painted on the body or stamped on the metal terminations, lower values sometimes conforming to a colour-code.

Variable resistors
Variable resistors are formed with a resistive track made usually
from a carbon composition mixture (although other types exist)
and a wiper arm, position of which may be varied along the track
and which acts as a pick-off point. Usually a shaft is joined to

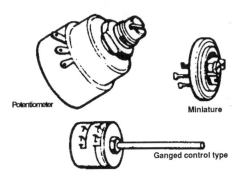

Potentiometer

Miniature

Ganged control type

the wiper arm, so that rotary movement of the shaft adjusts wiper
arm position. A special case is the pre-set variable resistor
whose wiper arm is adjusted by turning with a screwdriver-type
tool.

Variable resistors are used as rheostats or potential dividers
to vary current or voltage. Rheostats have one end of the
resistive track in circuit and also the wiper arm. Potentiometers,
on the other hand, have both ends of resistive track connected in
circuit, as well as the wiper arm. Main variable resistor types
are:

● carbon track (made with linear or logarithmic characteristics)

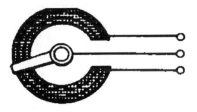

● wirewound (toroidal)

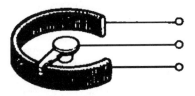

● wirewound (single loop)

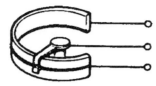

● tappe (switched change).

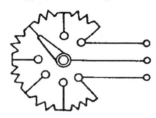

Identification

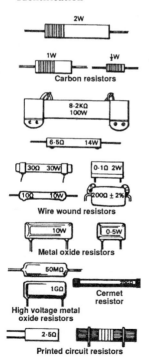

2W

1W   ¼W
Carbon resistors

8·2KΩ
100W

6·5Ω   14W

30Ω 30W   0·1Ω 2W

10Ω 10W   200Ω ± 2%
Wire wound resistors

10W   0·5W
Metal oxide resistors

50MΩ

1GΩ   Cermet resistor
High voltage metal oxide resistors

2·5Ω
Printed circuit resistors

Resistors are identified by their resistance value (ohms) and their power dissipating value (watts). Resistance value is either printed as a value, or coded as a colour-code on the body of a resistor. Wattage is generally indicated by component size, except in the case of large wattage components which may have the value printed.

Resistor colour-code
Resistors are colour-coded as specified in BS1852, by coloured
bands around the component body.

Colour	Significant figure	Decimal multiplier	Tolerance (%)
Black	0	1	
Brown	1	10	1
Red	2	100	2
Orange	3	1000	
Yellow	4	10000	
Green	5	100000	
Blue	6	1000000	
Violet	7	10000000	
Grey	8	100000000	
White	9	1000000000	
Gold		0.1	5
Silver		0.01	10
No colour			20

Most common is a four-band code where bands, numbered from
the end to which they are closest, are:

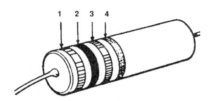

Band 1   first figure of resistance value
Band 2   second figure of resistance value
Band 3   number of noughts after first two figures
Band 4   percentage tolerance.

Used much less frequently is a five-band code in which first
three bands indicate value, fourth band indicates multiplier,
while the fifth indicates tolerance.

Printed value

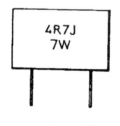

In this method figures and
letters are used to indicate
resistance, a letter indicating
the decimal multiplier(R = 0, K
= 1000, M = 1000000) taking
the place of a decimal point.
Tolerance is shown as a letter
which corresponds to a
percentage (F = 1, G = 2, J = 5,
K = 10, M = 20). Examples of
resistor markings are:

1Ω	=	1R0	3330Ω	=	3K3
4.7Ω	=	4R	15000Ω	=	15K
68Ω	=	68R	100000Ω	=	100K
100Ω	=	100R	2200000Ω	=	2M2
680Ω	=	680R	470000000Ω	=	47M

Preferred values

As exact values of fixed resistors are usually unnecessary only certain preferred values are made. This helps to reduce numbers of different values which need to be stocked. Values are selected in a number of series such that they give continuous overlapping ranges. Examples of preferred value series are:

● E12; a series with a 10% tolerance (1.0, 1.2, 1.5, 1.8, 2.2, 2.7, 3.3, 3.9, 4.7, 5.6, 6.8, 8.2, and multiples of 10)
● E25; a series with a 5% tolerance (those in the E12 series, as well as 1.1, 1.3, 1.6, 2.0, 2.4, 3.0, 3.6, 4.3, 5.1, 6.2, 7.5, 9.1 and multiples of 10).

Kirchhoff's laws

Ohm's law suffices for many calculations relating to electrical circuits but Kirchhoff's laws are used for more complex circuits.

First law

$$I = I_1 + I_2 + I_3$$

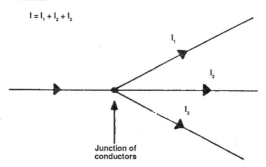

At any junction the sum of currents flowing towards the junction is equal to the sum of currents flowing away from it.

Second law

$$E = V_1 + V_2 + V_3$$

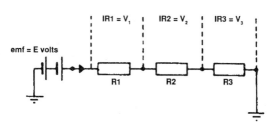

Algebraic sum of the products of current and resistance forming a closed path in a network is equal to the algebraic sum of electromotive forces in the path.

Resistors in series

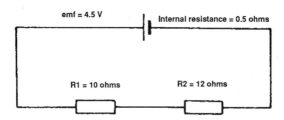

Features of a series circuit are:

● the same current passes through each resistor. It is not possible for the current through R1 and R2 to be different
● total circuit resistance is the sum of individual resistances
● sum of individual potential differences is equal to the applied emfs (Kirchhoff's second law)
● a break in one resistor opens the entire circuit.

Total resistance of resistors in serial: $R = R1 + R2 + ...$

Resistors in parallel

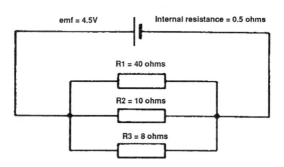

Features of a parallel circuit are:

● value of potential differences is the same for each resistor
● sum of currents through individual resistors is equal to total current (Kirchhoff's first law)
● value of combined or effective resistance is always less than that of the smallest resistance in the group
● total current divides in inverse proportion to branch resistances.

Total resistance of resistors in parallel:

$$\frac{1}{R} = \frac{1}{R1} + \frac{1}{R2} + ...$$

Potential divider networks

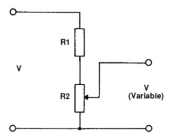

Provided current is not required to flow in the output, then for
a fixed output divider:

$$v = V \frac{R2}{R1 + R2}$$

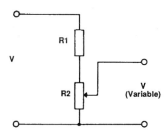

For a variable output divider, output is variable between limits
of $v = 0$ and:

$$v = V \frac{R2}{R1 + R2}$$

If current is allowed to flow in the output circuit, output voltage
is reduced, its value depending on load resistance. Potential
divider networks are used in many instruments, the most usual
being the standard workshop and temperature measuring instruments which incorporate a Wheatstone bridge circuit.

Special resistors

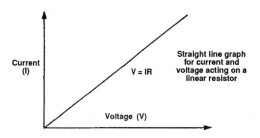

Straight line graph
for current and
voltage acting on a
linear resistor

$V = IR$

Current
(I)

Voltage (V)

So far only linear resistors have been considered. A linear resistor is one where voltage across it is directly proportional to current through it. Many non-linear resistors are used in electronics. These resistors do not follow the characteristic here.

Thermistor

A thermistor is a thermally sensitive resistor, comprising a semiconductive material whose resistance changes markedly with changes in temperature. There are two general groups:

● those with a negative temperature coefficient (ntc) whose resistances reduce with increasing temperature
● those with a positive temperature coefficient (ptc) whose resistances increase with increasing temperature (only over a limited temperature range).

Ntc types are made from oxides of nickel, manganese, copper, cobalt and other materials. They are commonly used for temperature control and measurement. Ptc types are based on barium titanate and are used mainly as protection devices to prevent damage in circuits which might experience large temperature rises (eg, motor overload protection circuits).

Thermistors are manufactured in rod, bead and disc forms with diameters as small as 0.015cm.

Photoconductive cell or light dependent resistor (LDR)
Resistance of certain semiconductors decreases as intensity of light falling on them increases. The effect is due to light energy setting free electrons from donor atoms in the semiconductor, so increasing its conductivity, and decreasing its electrical resistance. Although cadmium sulphide (cds) is the most common material used (as its spectral response closely matches that of the eye) other materials used include lead sulphide, lead selenide and indium antimonide.

One form of construction is shown, in which a film of semiconducting material is laid down on an insulating base, and electrodes are laid on its surface. The electrodes have a patterned shape to increase contact area and decrease resistance of the cell.

Metal electrodes on surface of cadmium sulphide

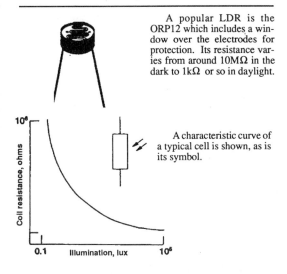

A popular LDR is the ORP12 which includes a window over the electrodes for protection. Its resistance varies from around 10MΩ in the dark to 1kΩ or so in daylight.

A characteristic curve of a typical cell is shown, as is its symbol.

Capacitors

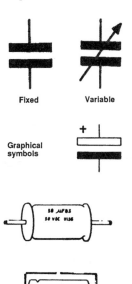

Fixed

Variable

Graphical symbols

Basic capacitor format consists of two metal plates, separated by an insulating layer which acts as a dielectric. This format allows the capacitor to store a charge. Capacitors fall into two main groups; electrolytic and non-electrolytic capacitors.

Insulating material of the electrolytic capacitor is formed by the action of an electric current upon aluminium conducting plates. Electrolytic capacitors are polarised ie, correct polarity must be applied to the plates, so terminals are marked positive and negative. If reverse polarity is applied the insulating layer is destroyed and the capacitor is short-circuited, destroying it and possibly other circuit components. Under such conditions a capacitor may explode. For the same reason electrolytic capacitors should never be connected across AC supplies.

Both types of capacitors are marked with working voltage (ie, maximum permissible voltage allowed across its terminals).

There are many types of non-electrolytic capacitors, including:

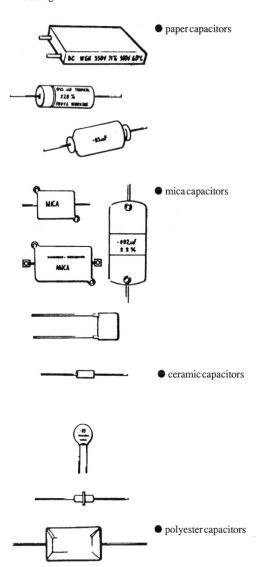

● paper capacitors

● mica capacitors

● ceramic capacitors

● polyester capacitors

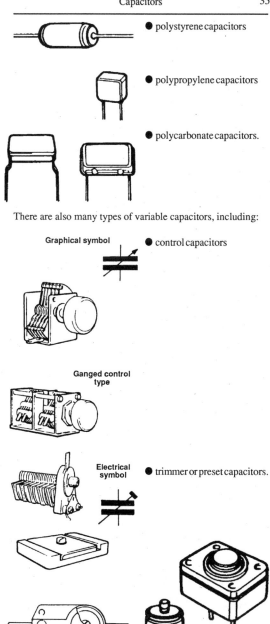

● polystyrene capacitors

● polypropylene capacitors

● polycarbonate capacitors.

There are also many types of variable capacitors, including:

Graphical symbol

● control capacitors

Ganged control type

Electrical symbol

● trimmer or preset capacitors.

Units of capacitance

Unit of capacitance is the farad (F). One volt applied to a capacitance of one farad maintains a charge of one coulomb. The farad is too large a unit for most practical purposes, so smaller units are more commonly used:

● the microfarad (μF) = one millionth of a farad
● the nanofarad (nF) = one thousand-millionth of a farad
● the picofarad (pF) = one million-millionth of a farad.

Capacitance value of a capacitor depends on:

● area of metal plates; increasing the effective area increases capacitance
● distance between plates; decreasing plate separation increases capacitance
● dielectric substance; various dielectrics are used to make many ranges of capacitor.

Capacitor colour-code

BS1852 defines capacitor colour-code.

Colour	Significant figure	Decimal multiplier	Tolerance (%)
Brown	1	10	1
Red	2	100	2
Orange	3	1000	
Yellow	4	10000	
Green	5	100000	
Blue	6	1000000	
Violet	7	10000000	
Grey	8	100000000	
White	9	1000000000	
Gold		0.1	5
Silver		0.01	10
No colour			20

Typically, though not always, these colours are marked in bands or spots on capacitors, in the following format:

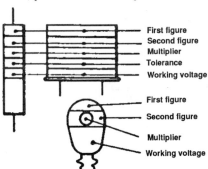

First figure
Second figure
Multiplier
Tolerance
Working voltage

First figure
Second figure
Multiplier
Working voltage

Band 1   first figure of the value
Band 2   second figure of the value
Band 3   number of noughts after the first two, to give value in picofarads
Band 4   tolerance
Band 5   maximum working voltage.

Capacitor maximum working voltage rating colour-code
Tantalum

white	3V
yellow	6.3V
black	10V
green	16V
blue	20V
grey	25V
pink	35V

Polycarbonate

red	250V
yellow	400V
blue	630V

Polyester

red	250V
yellow	400V

Printed value

Values are sometimes indicated by letters and numbers. Letters used to indicate the decimal multiplier are placed in the position of the decimal point.
These letters should conform to:

Letter code	Decimal multiplier
M	1
F	10
G	100
J	100000
K	1000000000

Sometimes a number is used as decimal multiplier, positioned after the value (usually in picofarads). Variations exist so, if in doubt, consult manufacturers' literature.

Capacitor use
Capacitors have many uses, including:

● allowing AC to pass while blocking DC
● filter elements in power supplies
● temporary charge holders
● integral parts of timing and tuning circuits.

Capacitors can be checked with a capacitance meter and it is also possible to quick-test them with an ohmmeter, though not as reliably. When checked with an ohmmeter, current output of the meter charges the capacitor quickly at first. Result is a definite low resistance which climbs towards infinity as the capacitor charges. This test cannot reveal serious defects, though it often reveals a shorted or open-circuit component. Low value capacitors cannot be checked with this method, however, because they charge so fast the meter cannot respond.

When replacing capacitors it is preferable to use exact valued replacements, as some circuits are designed around capacitors with special characteristics and so do not operate with a different type.

Capacitors in series

When capacitors are connected in series the effect is one of increasing plate separation so capacitance decreases. Total capacitance (C) of series-connected capacitors is given by:

$$\frac{1}{C} = \frac{1}{C1} + \frac{1}{C2} + \ldots$$

or for only two capacitors in series:

$$C = \frac{C1 \times C2}{C1 + C2}$$

Working voltage of the series-connected capacitors is the sum of the individual capacitors' working voltages. Current in the circuit is the same at all points. As a result smaller capacitors require a larger voltage to maintain a charge than larger capacitors.

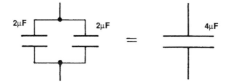

Capacitors in parallel

When capacitors are connected in parallel the effect is one of increasing plate area, so capacitance increases. Total capacitance of parallel-connected capacitors is given by:

$$C = C1 + C2 + \ldots$$

Working voltage of a parallel-connected capacitor circuit equals the working voltage of the capacitor with the lowest working voltage.

Effect of capacitance in a DC circuit

When a capacitor is connected to a DC supply the capacitor plates reach the supply voltage. If the supply is then disconnected, plates remain at the supply voltage; the capacitor is said to be charged. If the capacitor leads are then connected together through a load, the capacitor discharges.

A capacitor in a DC circuit only allows current to flow until fully charged; at which time current ceases. A capacitor is thus a barrier to flow of DC current.

RC time constant

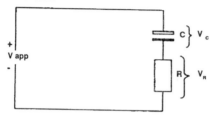

Operating frequency of much equipment is critical. If, say, television sweep frequency is not precisely what it should be, the picture rolls or tears. Computers, also, cannot function properly if main operating frequency is not constant.

In many cases frequency is determined and adjusted with an RC time constant: defined as time required for a capacitor to charge to 63.2% of applied voltage (Vapp), or time for current to fall to 36.8% of maximum. Mathematically, time constant is given by the expression:

t (in seconds) = R (in ohms) x C (in farads)

so the term RC is often known as the time constant, too. As resistance is always present in any capacitor circuit (either as a discrete resistance introduced into the circuit, or contained in the internal structure of a capacitor), it takes a definite time from application of a charging voltage to complete capacitor charge.

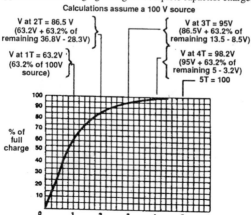

Calculations assume a 100 V source

V at 2T = 86.5 V
(63.2V + 63.2% of
remaining 36.8V - 28.3V)

V at 1T = 63.2V
(63.2% of 100V
source)

V at 3T = 95V
(86.5V + 63.2% of
remaining 13.5 - 8.5V)

V at 4T = 98.2V
(95V + 63.2% of
remaining 5 - 3.2V)

5T = 100

% of full charge

Capacitor charge increases in an exponential curve. In a time equal to 1t seconds a capacitor charges to 63.2% of applied voltage. In 2t seconds the capacitor charges a further 63.2% of the remaining voltage, or 86.5% of total applied voltage. In 3t seconds the capacitor charges to some 95% of total applied voltage. In 4t seconds this has increased to 98.2% of total, while time taken to full charge is generally regarded as 5t seconds.

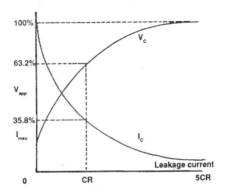

When the supply is first applied, initial voltage across the capacitor is zero while initial current is V_R/R, which at this time is Vapp/R. As charge on the capacitor increases, so voltage across the resistor decreases hence causing current to fall. This continues until charge eventually reaches applied voltage and circuit current ceases. This gives a charging graph for voltage Vc and current Ic.

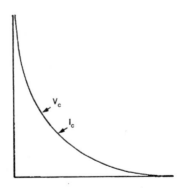

If the applied voltage is removed and the circuit terminals shorted together the capacitor discharges at the same rate, losing 63.2% of its charge per time constant. However, discharging current and voltage both fall with time, according to a discharge graph. The capacitor is considered to be fully discharged in 5t seconds.

Often a circuit is designed to have a different resistance on charge than on discharge, which means time constant for charge and discharge are different. In the example shown, a capacitor is charged through a relatively long time constant and discharged through a short one.

When voltage is first applied potential across the capacitor begins to rise from zero at a rate determined by values of resistor R and capacitor C. Before reaching a fully charged condition (at 5t) firing potential of the neon lamp is reached and the capacitor discharges through the very low resistance of the neon lamp. Discharge is thus much more rapid than charge. When charge falls to the switch-off potential of the lamp, the lamp becomes open circuit and the capacitor charges through the resistor again. This action is repeated as long as supply is present, producing the sawtooth waveform shown.

Another application of capacitor time constant is the simplified electronic photographic flash circuit shown. When switch S1 is closed capacitor C charges towards 15V. After one time constant (3000Ω x 0.0001F = 0.3 seconds) the capacitor has charged to 63.2% of 15V, or 9.5 volts. After 5t (1.5 seconds) the

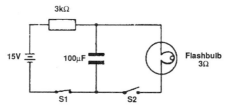

capacitor is fully charged. Now, when switch S1 is opened and switch S2 closed, the capacitor discharges through the flashbulb, which has a resistance of 3Ω. Naturally discharge time constant (3Ω x 0.0001F = 300 μs) is very much shorter than charge time constant. Rate of discharge is very important: faster the capacitor discharges, greater the current flow. The high discharge current thus produced is the reason for the very brief and intense flash of light from the flashbulb.

Residual charge

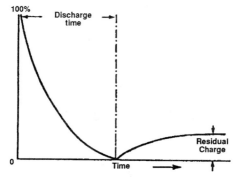

If the discharge curve is expanded along the time scale, beyond 5t seconds, the curve shown is obtained. This indicates a residual charge occurs across a capacitor after discharge. Residual charge occurs due to dielectric absorption and is important when handling capacitors which have been previously charged to high voltages; even though they have been discharged they may have a dangerous residual charge.

Capacitive reactance
Capacitors are used in many ways in both AC and DC circuits. However, they have a special effect on AC which they do not exhibit toward DC. This effect is known as capacitive reactance. It is essentially a reaction or opposition against being charged and discharged. It is in effect only when the capacitor is rapidly and constantly being charged and discharged and not when pure DC is applied.

Safety
Because of the high voltage which may be stored, capacitors of such a rating as to give rise to the possibility of electric shock after the supply is disconnected must be provided with a means of discharge. This is usually a high resistance path which allows charge to leak away. A voltage check can be used to indicate whether or not discharge is complete.

Inductors

Inductors (L) consist of a copper coil wound onto a core which may be magnetic. An inductor in a circuit opposes any change in current. Its action compares with that of a fly-wheel fitted to an engine; trying to oppose any change in engine speed. Design is influenced by operating frequency.

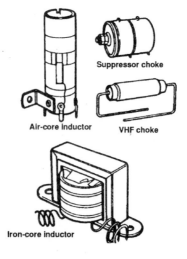

Suppressor choke

Air-core inductor   VHF choke

Iron-core inductor

Typical operating frequencies for various cores are:

Laminated iron 50 Hz
Mumetal   465 Hz
Air   over 1 MHz

Unit of inductance is the henry (H). Current increases at a rate
of one ampere per second when one volt is applied to a purely
inductive circuit of one henry.

Inductors in series
Total inductance is the sum of individual inductances. Thus,
total inductance L is given by:

$$L = L1 + L2 + ...$$

Inductors in parallel
Total inductance of parallel inductors can be found from the
expression:

$$\frac{1}{L} = \frac{1}{L1} + \frac{1}{L2}$$

Inductor connected to DC supply
Wire used for inductor coils possesses resistance, and this limits
DC current which flows for a given applied voltage. When an
inductor with an inductance of L henrys and resistance R ohms

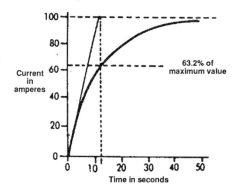

is connected to a DC supply of V volts, current rises exponentially
to a value of V/R amperes. Rate of change of current decreases
as current rises. If the initial rate of change could be maintained,
however, current reaches its maximum in t seconds, and this
time is called the circuit's time constant. In practice, current
reaches 63.2% of its maximum value in time t seconds. For an
inductor, time constant is given by the expression:

$$t = \frac{L}{R}$$

Inductive reactance

Inductive reactance is an inductor's effective resistance to current flow. It is directly proportional to inductance size and current frequency. Symbol for reactance is X and, as symbol for inductance is L, inductive reactance is X_L, measured in ohms.

Impedance

When an inductor is connected to a DC supply, maximum current is not limited, merely the time taken to reach maximum current is extended. When an inductor is connected to an AC supply, on the other hand, current is reduced. This opposition to AC current is known as impedance (Z) and is given by the expression:

$$Z = \frac{V}{I}$$

Identification

An inductor usually has inductance, resistance and impedance ratings printed on it. Colour-coding may be used on tubular insulated chokes. Where an inductor has several leads or terminals, identification is alphanumeric or colour-coding.

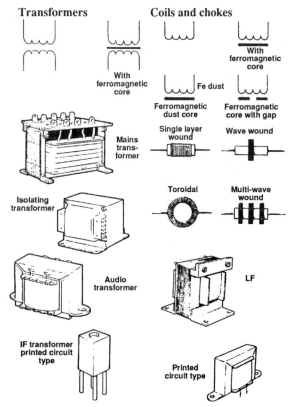

Transformers

With ferromagnetic core

Mains transformer

Isolating transformer

Audio transformer

IF transformer printed circuit type

Coils and chokes

With ferromagnetic core

Fe dust

Ferromagnetic dust core

Ferromagnetic core with gap

Single layer wound

Wave wound

Toroidal

Multi-wave wound

LF

Printed circuit type

Electric motors

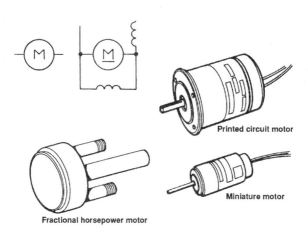

Printed circuit motor

Fractional horsepower motor

Miniature motor

Relays

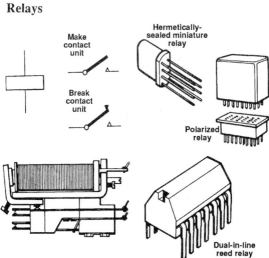

Make contact unit

Break contact unit

Hermetically-sealed miniature relay

Polarized relay

Dual-in-line reed relay

Fuses and circuit breakers

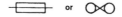

Fuses

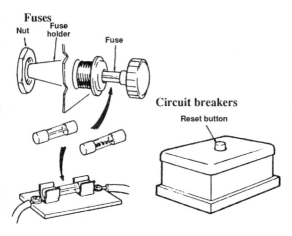

Circuit breakers

Reset button

Generators

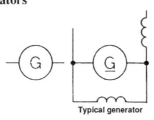

Typical generator

Synchros

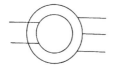

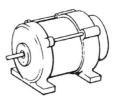

Switches

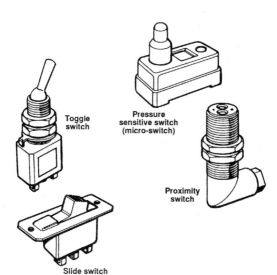

Toggle switch

Pressure sensitive switch (micro-switch)

Proximity switch

Slide switch

Semiconductors

A semiconductor is a material which is an insulator at low
absolute temperature, while at room temperature it has a conductivity between that of insulators and conductors. It also
possesses other unique properties displayed by neither insulators nor conductors. Application of heat increases electrical
conductivity, but this might have a cumulative effect which
destroys the material so must be avoided. Electrical conductivity is increased by adding a small amount of impurity to the pure
semiconductor material, in a process called doping. Some
semiconductors were used as rectifiers for many years before
the introduction of this doping technique.

Copper oxide and selenium rectifiers

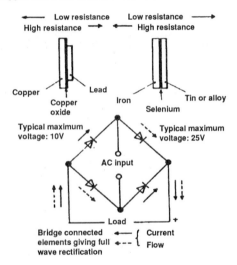

Construction of single rectifier elements is shown. Both types
have the same principle: a much higher resistance to current
flow in one direction than in the other. Four rectifier elements
are often bridged together to allow full wave rectification of an
applied AC voltage.

Structure of semiconductors

Silicon is now the most widely used semiconductor. All
semiconductors, however, have one common feature: the outer
shell in each atom contains four electrons (called valence
electrons) which are loosely held. Valence electrons largely
determine the chemical properties of the material and the
manner in which the atoms react with the atoms of other
materials.

Lattice diagram - pure semiconductor

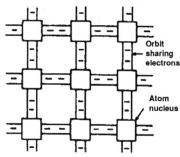

Crystal structure of pure germanium and silicon. For
simplicity, four outer electrons of each atom only are shown

It is convention to represent pure crystalline semiconductor
atoms by showing the four valence electrons as in the lattice
diagram shown. In this formation a bond exists between the
orbit-sharing electrons of adjacent atoms. Because of the lack
of loosely held electrons, the material in this crystalline state is
an insulator at low temperatures.

Production of N-type material
Atoms of antimony, arsenic and phosphorus have five electrons
in their outer shell. Doping pure silicon by adding small
amounts of these three impurities gives a new lattice arrangement, known as an N-type semi-conductor. Resultant material

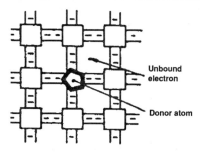

is still electronically neutral, but there is no place in the tightly
bound structure for the fifth electron in each donor atom. These
'spare' electrons may be moved by application of a small
electrical force; the material thus becoming an effective conductor.

Production of P-type material
Introduction of indium, boron or gallium as an impurity creates
a different molecular structure. These impurities are doping
atoms which have only three electrons in their outer shell,
creating a P-type semiconductor. Again the material is electronically neutral, but each doping atom is an acceptor which

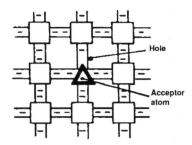

requires an electron to complete the binding force. This leaves
an electron deficiency, called a hole, in the vicinity of each atom.
A small applied emf causes electrons to jump from hole to hole
in the direction of negative pole to positive pole, with the result
that holes appear to travel from positive pole to negative pole.

Both holes and electrons in semiconductor materials are
known as charge carriers.

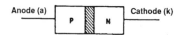

Junction of N- and P-type material

When a slice of semiconductor material is doped with different
impurities at each end, one end becomes N-type and the other
end P-type. The two zones meet in the middle to form a PN
junction. This semiconductor slice displays the characteristic of
a diode.   When a voltage is applied such that the P-type
semiconductor (the anode — labelled a) is positive and the N-

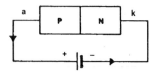

Conducting (forward biased)

type semiconductor (the cathode — labelled k) is negative,
excess electrons are attracted from the cathode to flow into the
anode; holes from the anode being swept into the cathode. Thus
a current flow is established. In this situation the junction is said

to be forward-biased and the current flow is known as forward current.

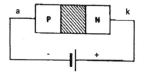

Non-conducting (reverse biased)

If polarities are reversed, on the other hand, electrons are attracted from the cathode and holes are attracted from the anode by the source of potential, leaving the junction void of holes and electrons, therefore unable to conduct. The junction only passes current in one direction. Ideally, when reverse-biased, current flow should cease. In practice the working temperature produces sufficient heat energy to break a few bonds within the junction so a small reverse current flows.

Diodes

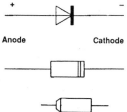

+ Anode   − Cathode

The semiconductor or PN junction diode

Semiconductor diodes belong to a family of devices known as solid-state. They are constructed of solid material as compared with vacuum tubes (also called

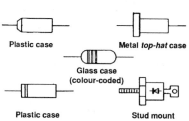

Plastic case

Metal *top-hat* case

Glass case (colour-coded)

Plastic case

Stud mount

valves) which require a vacuum to operate. Symbol for a diode is given. If the symbol is considered as an arrow, then this arrow indicates direction of conventional current flow — from positive to negative. Thus the major function of the diode is graphically represented: it presents a low resistance to current flow in one direction, while a high resistance is presented in the other direction.

Positive connection of a diode is called the anode, while the negative connection is the cathode. There is usually a bar at one end of a diode to represent the cathode.

Diode case types

Many case types of diode are available, capable of performing many tasks from rectifying AC in the order of hundreds of amperes, to high-speed logic switching. Amount of energy handled by a diode affects the way its manufacturer needs to encapsulate it. A selection of typical diode case types is given.

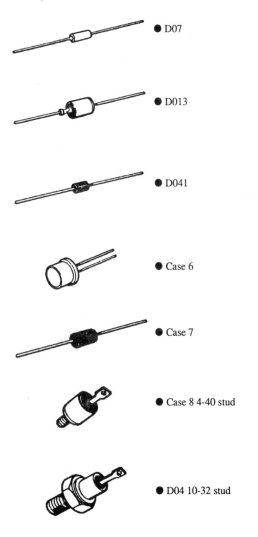

● D07

● D013

● D041

● Case 6

● Case 7

● Case 8 4-40 stud

● D04 10-32 stud

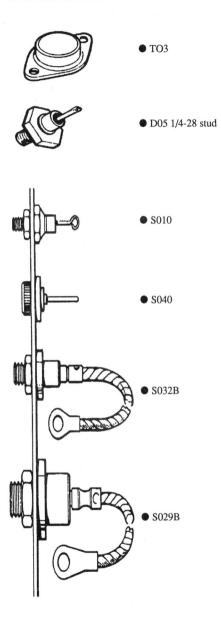

● TO3

● D05 1/4-28 stud

● S010

● S040

● S032B

● S029B

Use of junction diodes as rectifiers

Junction diodes are used in the following ways.

Half wave rectifiers
The diode conducts during positive half cycles, the arrow head indicating direction of conventional current when a load is connected.

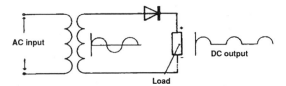

Full wave rectifiers
Two diodes connected to a centre-tapped transformer winding allow full wave rectification of the transformer's AC output. During one half cycle the top end of the winding is positive and the bottom is negative, both with respect to the centre tap.

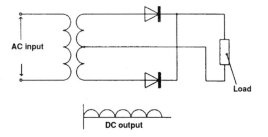

During the next half cycle the condition is reversed. Each diode conducts during the half cycle during which it is forward-biased.

Full wave rectifiers, bridge-connected

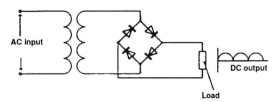

Two diodes in the bridge conduct during each half cycle, allowing full wave rectification from a transformer which is not centre-tapped.

Smoothing

Output of a rectifier is pulsating, and a steadier output is obtained with a smoothing capacitor.

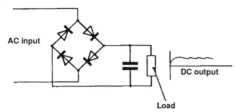

Voltage doubler circuit

A voltage doubler circuit produces an increased DC output voltage for a given AC supply voltage. In a basic circuit capacitor C1 charges on one half cycle, while capacitor C2 charges on the other half cycle. As the two capacitors are

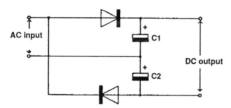

connected in series, DC output voltage is approximately twice the peak value of the AC supply voltage. Capacitors C1 and C2 must be capable of storing sufficient energy as required by the load.

Point contact diodes

A point contact diode consists basically of a sharp-ended wire pressing onto a thin wafer of semiconductor. After assembly a current in the form of a pulse is passed through the diode, forming a tiny PN junction.

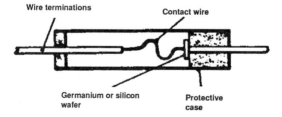

Because a point contact diode's internal capacitance is very small it is useful in high frequency applications.

Photodiodes

Photodiodes use the fact a reverse-biased diode can never totally prevent current. In the photodiode, light energy impinging on

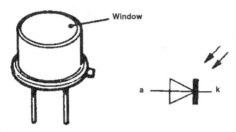

the PN junction is used to break bonds in the semiconductor, hence allowing a leakage current to flow which increases proportionally with incident light. Speed of response to changes in light is very high, typically 250 ns.

Typical applications include:

● high-speed counters for computer punch card readers
● photometers for measuring light intensity
● modulated light detectors in optical communications.

Light emitting diodes

Light emitting diodes (LEDs) are made from semiconductor crystals of indium phosphide or gallium arsenide and are housed in a translucent package. The LED is electrically similar to an

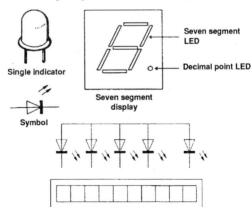

ordinary diode but, when forward-biased, light is emitted as a result of electron/hole recombinations in the junction. Colour of light output is dependent upon the material used. Infra-red, red, green, yellow and amber are common.

When used as simple indicators LEDs provide a long-life alternative to filament lamps with no heat dissipation problems. By modulating forward current (and hence light output) LEDs are used in telecommunications systems to provide electrical isolation between systems. When used in arrays, LEDs can produce bargraph indicators and seven-segment displays.

Zener diodes

A zener diode is a special-purpose diode used for voltage regulation (keeping voltage constant). It has the special feature of conducting in reverse-biased mode when applied voltage is

over a certain voltage. Underneath the voltage, however, the diode prevents conduction as normal. This feature is used in voltage regulation circuits to maintain a constant voltage output for a varying voltage input. Another use is to 'clip' or make squarewaves out of applied sinewaves. Zeners are found in almost all electronic devices.

Transistors

Transistors are three-terminal semiconductor devices. Most common type of transistor is the bipolar transistor. Terminals are called:

● base
● emitter
● collector.

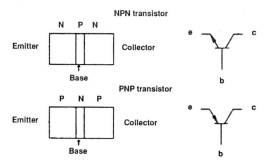

Bipolar transistors are usually made of silicon, although germanium is sometimes used, and consist of two semiconductor PN junctions following either an NPN or a PNP format. Arrow on a symbol's emitter indicates direction of conventional current flow when the emitter-base junction is forward-biased.

Construction of junction transistors
There are two methods used in formation of transistor junctions.

Grown junctions
Large semiconductor crystals are produced by withdrawing a seed crystal at a controlled rate from a molten mass. During the operation doping at appropriate intervals to give alternate N- and P-type material forms the required junctions. Transistors

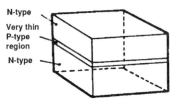

are then made from thin slices cut from the resulting crystal.

Fused junctions
Small pellets of indium are fused onto each side of a thin slice of N-type material. Fusing produces regions of P-type material; giving PNP junctions.

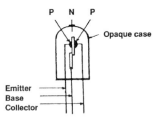

Semiconductor junctions are light-sensitive and the finished assembly is housed in an opaque case. Due regard should be paid to precautions when handling transistors, described later.

Transistor types

A very wide range of transistors is available, capable of performing many tasks from amplifying very small signals and switching at ultra-high frequency to regulating current at very high power. Encapsulation of transistors greatly affects the way they are used. A selection of common transistor package types is shown overleaf.

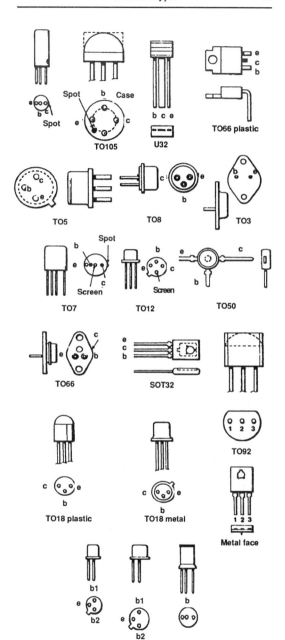

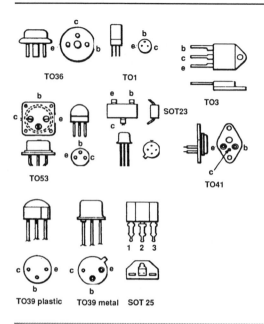

TO36   TO1

TO53   TO41

                                  TO3

TO39 plastic   TO39 metal   SOT 25

Transistor configurations

Transistor amplifiers are connected in three configurations:

● common base (base common to input and output)

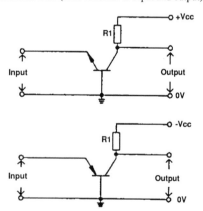

● common emitter (emitter common to input and output)

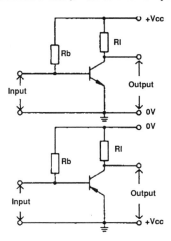

● common collector (collector common to input and output).

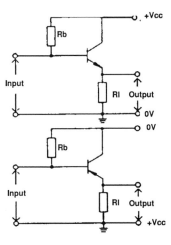

Transistor as an amplifier

A change in current at the input, in the form of an applied signal, causes a change in output current. How one current change is related to the other depends on which configuration is used.

Common base configuration

Input signal is applied to the emitter, while output is obtained from the collector. In this configuration the collector current change is 95 to 98% of the emitter current change. Current gain

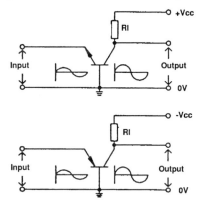

(given the symbol h_{fb}) is calculated from:

$$h_{fb} = \frac{\text{output current change}}{\text{input current change}}$$

Current gain of a transistor connected as common base is thus always less than unity.

Because the emitter-base junction is forward-biased, its junction resistance is very low. However, because the collector-base junction is reverse-biased, its junction resistance is high. When current changes take place through these different resistances, voltage amplification occurs. Input and output signals are in phase.

Common emitter configuration

Input signal is applied to the base in this circuit, while output is from the collector. A small current change in the base circuit produces a much larger current change in the collector circuit, so current amplification results. Current gain is calculated from:

$$h_{fe} = \frac{\text{change in collector current}}{\text{change in base current}}$$

Input resistance is high, while output resistance is low. Voltage inversion occurs between input and output.

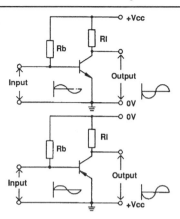

Common collector configuration

Here input is to the transistor base, while output is from the emitter. Current gain is similar to the common emitter configuration, while input resistance is high, output resistance is low, and input and output voltages are in phase.

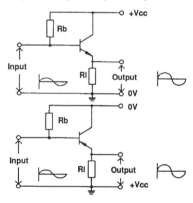

A summary of the three transistor configurations follows:

	Common base	Common emitter	Common collector
Input resistance	very low	medium	high
Output resistance	high	medium	low
Phase inversion	no	yes	no
Current gain	less than 1	more than 1	more than 1
Typical application	high frequency amplifier	multi-stage amplifier	impedance matching

Feedback

Feedback is a technique which transfers part of a circuit's output back to its input. Application of feedback affects a circuit's gain and stability.

Negative feedback

In this form, sometimes known as degenerative feedback, current fed back is 180° out-of-phase with the input and so decreases gain. Typically, common emitter configured transistors are used when negative feedback is required, as normal operation involves phase inversion.

Output waveforms of many amplifier circuits are distorted to some degree. With negative feedback, however, the effect of output superimposed on the input reduces distortion in the output. Negative feedback also reduces overall gain and increases stability.

Positive feedback

Also sometimes known as regenerative feedback, a fraction of the output is fed back in phase to the input. Application of positive feedback increases an amplifier's gain and output distortion. If feedback is sufficient the amplifer becomes unstable and oscillates.

Stabilisation of transistor amplifiers

Transistor leakage current, which increases with temperature, can cause instability in an amplifier. Stability is improved with negative feedback. Feedback is applied by resistor Rfb. With the amplifier in a stable starting condition, a rise in temperature (due to normal operation) causes a rise in leakage current, so the collector current is greater. Potential difference across resistor

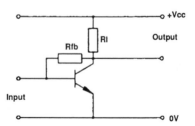

Rl rises so the collector voltage falls. Base current, which depends upon collector voltage (via Rfb) falls so collector current decreases. Thus the effect of temperature change is minimised.

If negative feedback is undesirable (because of its effect on amplifier performance) stabilisation can be achieved another way. Here, base bias voltage is obtained with a potential divider

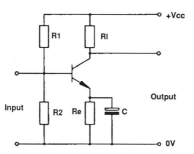

network comprising resistors R1 and R2. With this network the base voltage is held steady in the absence of a signal so bias current is determined with emitter resistor Re. Decoupling, using capacitor C, is usually required.

A rise in leakage current due to, say, a temperature rise, causes an increase in collector current and emitter current. The larger emitter current creates a larger potential difference across emitter resistor Re, which causes the emitter voltage to rise and so the forward bias of the base/emitter junction decreases. Current to the base is reduced, causing a reduction of collector and emitter current, counteracting effects of change in leakage current.

Multi-stage amplifiers

Multi-stage amplifiers are used in applications when gain of a single stage is insufficient for a purpose. As input and output resistances of transistors vary, depending basically on configuration, it is essential impedance matching techniques are used when coupling stages. Impedance matching means ensuring output impedance of one stage is similar to input impedance of a following stage. This ensures maximum transfer of power between stages.

Multi-stage AC amplifiers
Two methods of connecting AC amplifier stages are common. They are both reactance coupling techniques are so are not suitable for DC amplifiers. Methods are:

● resistance-capacitance (RC) coupling. This method is inexpensive, simple and the capacitor occupies a small space. Because of these features it is widely used in spite of the fact some loss in amplification usually occurs due to differences between input and output resistances of stages

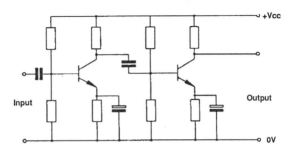

● transformer coupling. This form of coupling is efficient because impedances of primary and secondary windings of transformers can be closely matched to inputs and outputs of stages. Winding impedances, however, change with frequency, so amplifier gain varies with frequency. Transformer coupling is more expensive than RC coupling, and transformers are relatively heavy and bulky.

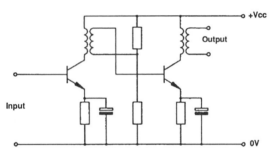

Multi-stage DC amplifiers
Reactance coupling cannot be used in a DC amplifier because both capacitors and transformers block DC. A multi-stage DC amplifier uses transistors with characteristics allowing them to be directly coupled — the output of one stage is directly connected to the input of the next.

This type of circuit suffers from serious drawbacks, however, most important being zero drift. This is caused because any change in leakage current in one stage is treated as a signal by the following stage, and amplified. Thus a substantial change in output can occur with a relatively small temperature change.

To overcome this problem the circuit may include two transistors TR1 and TR2 in an arrangement commonly known as a long-tailed pair. These are selected with identical characteristics and mounted together on a heatsink to maintain equal temperature. The transistors are connected in common emitter configuration and base bias is provided by identical potential divider circuits (Ra and Rb, Rc and Rd). One emitter resistor Re is shared by the transistors and a potentiometer Rbal is used to balance the two collector voltages.

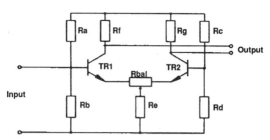

Input to the circuit is to the base of transistor TR1, while the base of transistor TR2 is held at a steady reference voltage by its potential divider network. A change in input voltage creates a corresponding change in output voltage at the collector of transistor TR1. Collector voltage of transistor TR2, however, stays the same. Output is thus the difference between the two collector voltages.

A rise in leakage current in one transistor is accompanied by an equal rise in the other, thus both collector voltages rise equally for a change in leakage current and no change in output occurs.

Transistor as a switch

Transistors are used as solid-state switches, where loads can be switched on and off without moving contacts. Principle is shown using a two-position mechanical switch — in practice the required change in voltage is produced by a preceding circuit.

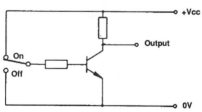

With the switch in the on position the emitter/base junction is forward-biased and the transistor conducts (ie, there is a large collector current). With the switch in the off position the emitter/base junction is reverse-biased and the transistor does not conduct (there is no collector current although, in practice, there is always a small leakage current through the collector). Advantages of transistors as switches include:

● no moving parts to wear out
● very high speed switching can be carried out
● transistors are cheap and, with large scale integration (LSI), integrated circuits with hundreds of thousands of transistors are common.

Oscillators

A charged capacitor in parallel with an inductor forms an oscillatory circuit, the capacitor alternately discharging and charging. Frequency of oscillation f of such an oscillator is given by:

$$f = \frac{1}{2\pi LC}$$

where L is inductance in henrys, and C is capacitance in farads. Such a circuit is often called an LC circuit.

Oscillations in practical circuits are damped due to resistive losses so, for sustained oscillation, these losses must be replaced.

One method of replacing losses is the tuned collector oscillator. Here a resistive network R1 and R2 provides sufficient bias to cause collector current flow. Consequent build-up of

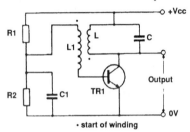

· start of winding

current through inductor L sets up a changing magnetic field which links with inductor L1, inducing an emf into the base circuit. If the emf is in phase with the base it creates a form of positive feedback, which makes up the losses of the LC circuit.

Multivibrators

A multivibrator generally comprises a two-stage transistor amplifier, with both stages connected in common emitter configuration. A large amount of positive feedback is employed. Multivibrators are used:

● as oscillators
● to produce pulses of square waveform
● as fact-acting switches.

Astable multivibrator

This is often referred to as a free-running multivibrator and it produce a stream of squarewave pulses. Phase inversion takes place over each of the two amplifier stages and positive feedback results because the second stage output is fed back to the first stage input. This feedback drives the amplifier into sustained oscillation. In this condition, one transistor is conducting (ie, switched on) while the other is switched off, and this condition reverses alternately.

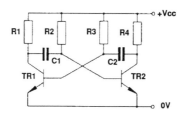

When transistor TR1 is switched on and transistor TR2 is off, capacitor C2 charges via resistor R4 and the fairly low resistance of the base of transistor TR1. Meanwhile capacitor C1 charges via resistor R2 and transistor TR1. As the base voltage of transistor TR2 reaches 0.6V (due to capacitor C1 charging), the transistor switches on and its collector voltage falls to almost 0V. This causes the junction of capacitor C2 and resistor R3 to swing negative, reverse-biasing the base/emitter junction of transistor TR1 and turning the transistor off. Process is now reversed.

Switching between the two transistors continues because the circuit is not stable in either state. Oscillation frequency is dependent on values of capacitors C1 and C2 and resistors R2 and R3.

Monostable multivibrator
This uses a similar two-stage amplifier circuit, in which one stable state exists when one transistor is on and the other off. However, a pulse of correct polarity applied to either transistor base drives the multivibrator into its unstable state. After a time the circuit switches back to its stable state.

In the basic monostable multivibrator transistor TR1 is held off by the voltage derived from potential divider network of

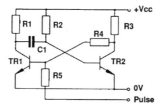

resistors R4 and R5. Meanwhile transistor TR2 is held on with base current via resistor R2.

If transistor TR1 is switched by application of a pulse, its collector voltage falls to almost 0V. This drives the junction of capacitor C1 and resistor R2 negative, which turns off transistor TR2.

Immediately capacitor C1 starts to charge via resistor R2 and transistor TR1. When its junction with resistor R2 rises to about 0.6V transistor TR2 turns on again. Feedback action rapidly turns transistor TR1 off, returning the circuit to its stable state.

Bistable multivibrator

This multivibrator has two stable states, each occurring when one transistor is off and the other is on. A pulse drives the multivibrator from one state to the other.

In the basic circuit, with transistor TR1 on and transistor TR2 off, the circuit is stable because transistor TR1 is held on by the collector voltage of transistor TR2, and transistor TR2 is held off by the collector voltage of transistor TR1

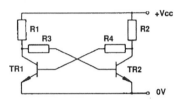

A negative pulse to the base of transistor TR1, however, turns the transistor off so its collector voltage increases turning transistor TR2 on. The circuit is stable again. A second pulse to the base of transistor TR2 returns the circuit to its first stable state.

Schmitt trigger

The Schmitt trigger is a switching circuit, capable of carrying out its operations at very high speed.

Resistor values are such that, before application of any input voltage, transistor TR1 is off while transistor TR2 is on.

As the input voltage rises above a triggering level voltage, transistor TR1 switches on which, in turn, switches transistor TR2 off.

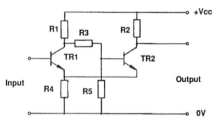

When the input voltage returns to a voltage lower than the triggering level voltage the transistors switch back to their original states.

Change in voltage level required to switch the circuit is very small, so the output of a Schmitt trigger often has a square waveform. Indeed, it is frequently used to produce a squarewave output from a sinewave input.

Other transistors

The unijunction transistor (UJT)
This pattern of transistor is made from either N-type or P-type material, to which a single P-N junction is made.

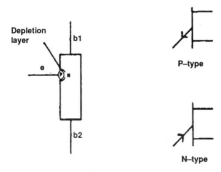

Depletion layer

b1

e

b2

P–type

N–type

A unijunction transistor has two bases, b1 and b2, with a typical resistance between the two of 10000 Ω. The third connection is called the emitter e. When a voltage is applied across the two bases (with base b2 positive with respect to base b1) the semiconductor strip effectively acts as a potential divider, and the emitter voltage depends on the emitter position. In this condition the emitter junction is reverse-biased.

If a sufficiently high positive voltage is applied to the emitter, the junction becomes forward-biased and emitter current (in the form of hole charge carriers) flows towards base b1. Presence of holes between emitter and base b1 means there is an increase in electrons between emitter and base b2. This region conducts more and so resistance is less. The unijunction thus displays a negative resistance characteristic.

Unijunction transistors are used mainly as pulse generators in thyristor circuits, and in some oscillator circuits.

Field effect transistor (FET)
These differ from conventional bipolar transistors as they have very high input impedances. Like bipolar transistors they have three terminals, but the field effect transistor terminals are called:

● drain
● gate
● source.

Input signal of a field effect transistor controls current flow through a region called a channel in the transistor. In a junction gate field effect transistor (JUGFET) this current control is via a PN junction, while in an insulated gate field effect transistor (IGFET) a capacitive connection controls current.

Because of high input impedances, special care must be taken to avoid damage. With some FET types it is important to:

● wrap a tinned copper wire around the leads when handling or soldering
● use a low voltage iron for soldering.

Junction gate field effect transistor (JUGFET)

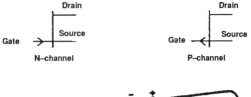

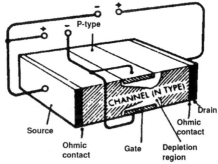

These transistors are made from either N-type or P-type semiconductor to which gates of opposite type semiconductor are attached.

When a voltage is applied to the end connections, electrons flow through the channel. If the gate connection is made negative, the PN junctions are reverse-biased and a region devoid of charge carriers is created near the junctions. These regions are called depletion regions.

As the gate becomes more negative, the depletion regions grow until finally conduction through the channel ceases. Gate voltage required to reduce current flow to zero is called pinch-off voltage.

Advantages of the junction gate field effect transistor are its high input resistance and a small capacitance.

Insulated gate field effect transistor (IGFET)
Substrate of an insulated gate field effect transistor can be either N-type or P-type material.

In a P-type substrate, two parallel N-type regions form source and drain junctions of the IGFET. A capacitive gate is formed with a layer of dielectric material and an outer layer of aluminium. The dielectric material is often silicon oxide and when this material is used the transistor is known as a metal oxide semiconductor transistor (MOST or MOSFET).

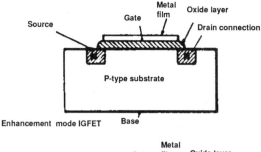

Enhancement mode IGFET

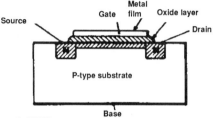

Depletion mode IGFET

If a voltage is applied across the source and drain connections a small leakage current flows as one of the PN junctions is reverse-biased. If consequently a voltage is applied across the gate and base, with gate positive, then a movement of positive and negative charge carriers takes place. Negative carriers are attracted to the surface of the substrate between the two N-type regions, while positive carriers are driven into the substrate.

Thus a conducting channel is formed, amount of conduction depending on gate-to-base voltage. Channel formed by negative charge carriers is called an inversion layer, because conduction changes from P-type to N-type when the positive gate voltage is applied. An increase in positive gate voltage produces an increase in current flow and this type of action is called an enhancement mode.

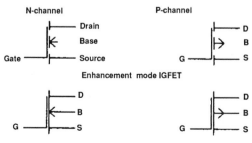

Opposite action to enhancement mode is obtained when transistor construction is modified. If, during manufacture, a thin N-type layer is formed on the substrate surface between the two N-type regions then, in absence of gate voltage, there is a conducting channel between the regions. When a negative voltage is applied to the gate, conductivity of the channel is reduced. This action is called depletion mode.

Insulated gate field effect transistors are used in applications where high input impedances are required, because their input resistance is very high and capacitance is extremely small.

Thyristors

The thyristor, or silicon controlled rectifier (SCR), is a four-layer PNPN device, with an anode, cathode and a controlling gate. It may be likened to two diodes in series, forming three junctions: A, B, C.

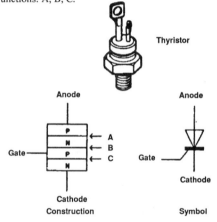

Thyristor

Construction

Symbol

Operating conditions
Thyristors have three main operating conditions:

● forward blocking; in which only a small leakage current flows, with anode positive with respect to cathode
● reverse blocking; in which device characteristics are similar to a reverse-biased diode
● forward conducting; in which anode to cathode resistance falls to a very low value, so a large current is passed at a small voltage drop.

A thyristor can become forward conducting if its gate is made positive with respect to its cathode. Gate current required to cause forward conduction is small and may only be for a short period. Once entering forward conduction the thyristor does not again block current until forward current is reduced below a low value called the holding current.

A triac is a two-way thyristor and conduction is controllable in both directions. Its advantage is it takes place of two thyristors in full-wave circuit.

Triac

Opto-electronics

Apart from optical photodiode and light emitting semiconductor devices, there are many other optical devices.

Signal lamps
These are very small tungsten filament lamps, often encapsulated in a lamp holder. They are generally used in low voltage installations where current required to drive the lamp is not a problem.

Neon
These devices are available in a variety of shapes and sizes. They are frequently used as mains supply indicators. Consuming much less current than signal lamps they can be used over a wide range of voltages, with a suitable series resistor.

Filament lamps
These are similar to signal lamps, but physically separate from the lamp holder. Current consumption is quite high.

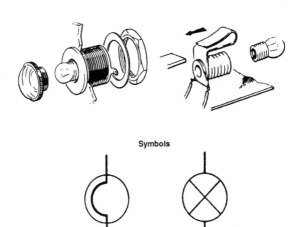

Symbols

Illuminating lamp   Indicating lamp

Thyristors are widely used in AC power handling circuits, such as control of motor speed or lamp brightness. Two control circuit types are common. These, and resultant relationships between gate, anode and load voltages, are:

● phase control; in which the thyristor conducts for part of every positive half cycle. It is used for control of fairly low power equipment, such as small motors and low wattage lamps, and has a disadvantage of causing power supply current surges which, in turn, create radio frequency interference (RFI)

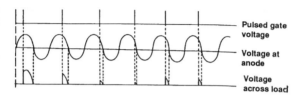

● burst triggering; in which the thyristor is switched on for a number of positive half cycles then switched off, in a repeating manner. Burst triggering causes much less radio frequency interference as the thyristor switches on when anode voltage is zero, so current surges in the power supply are eliminated. One disadvantage, however, is time between power on and power off. This usually limits use of burst triggering to loads with large time constants (large motors or heaters, say).

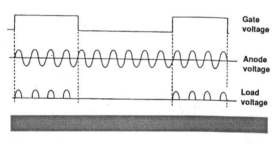

Diacs and triacs

Although used as common names for certain types of thyristors, the terms diac and triac are trade names of General Electric of America.

Diac

The diac is a diode version of the basic thyristor, and features no gate electrode. A voltage across the device must be greater than the triggering voltage for current to flow. Diacs are frequently used in gate circuits of thyristors to provide very sharp and defined triggering.

Liquid crystal displays (LCD)

Consuming very little current, these devices are ideal for battery-powered operation. They have black segments on a light grey background, and can be as simple as one digit or as complex as a whole computer screen display. They do not emit light, so cannot be read directly in low ambient light surroundings. In applications where such surroundings may occur, a backlight is provided. Such a backlight, however, defeats the extremely low current consumption (typical 5 µA) of the device.

Opto-isolators

These are devices used to electrically isolate parts of a system by combining a light emitter and receiver in one package. Output of one part of a system drives the light emitter, while the light receiver picks up the light converting it back to an electrical signal for the subsequent part of the system. Thus the only connection between the two parts is a non-electrically-conducting light signal. Opto-isolators are useful in applications where small currents (say, digital signals on a computer data bus) control large currents (for pumps, compressors and so on). Opto-isolators are usually packaged in a dual-in-line integrated circuit format, with a number of isolators in one package. They typically feature a light emitting diode device input, while output device is of a number of device-types:

- photodiode
- phototransistor
- optically coupled triac driver
- photo-darlington driver.

Fibre optics

This is a method of transmitting signals in the form of light along waveguides known as optical fibres. Although many varieties exist, there are basically two types of fibre:

- polymer; a low cost method of transmission used for short distance communications
- glass; higher fibre cost, although with much superior transmission capabilities. Used for long distance, high-bandwidth communications.

There are several forms of fibre within each type, grouped according to refractive indices and each having a different transmission characteristic, but all operate to the same principle.

Generally, single fibres are not used, instead many are combined together into a cable.

Integrated circuits

A complete electronic circuit containing transistors, diodes, resistors and capacitors, together with interconnections can be manufactured on a silicon chip of just a few millimetres square. The example shows such a circuit containing 34 transistors.

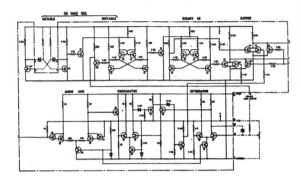

This is a highly magnified
picture of the silicon chip
embodying the circuit.

● This is actual chip size.

Many such circuits, after packaging in a larger body to
protect the chip and allow connections to be made to it, may be
soldered onto a printed circuit board.

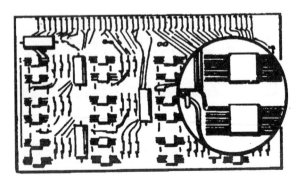

Integrated circuit (IC) development is the most important advance made in electronics. With its use, economic, compact and reliable systems (from watches to large digital computers) are now taken for granted as everyday items.

Historically, integrated circuits originate from large hybrid circuits (which, in turn, are a development of the printed circuit board technique) in which numerous discrete components are potted inside a single enclosure. Today, large scale integration (LSI) makes possible whole systems in a single integrated circuit package, at very low cost compared with a discrete component alternative.

The dual-in-line (DIL) package is the most common integrated circuit type, comprising a plastic or ceramic bar in which the integrated circuit itself is embedded. Connecting leads or 'pins' are brought out from the integrated circuit along the package sides, hence the term *dual-in-line*.

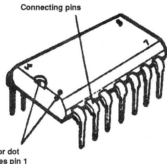

Connecting pins

Notch or dot identifies pin 1

Number of pins depends on complexity and purpose of the integrated circuit. Many packages have just 4, 7 or 8 pins on each side, making 8- 14- or 16-pin DILs. For clock, calculator or simple microprocessor packages 24- and 40-pin DILs are common. More complex integrated circuits require many more pins, however.

Two major categories of integrated circuits exist:

● linear, operating on analogue signals. Typical linear integrated circuits are operational amplifiers and timers. These are fairly conventional devices used regularly in audio, control and video circuits
● digital, operating on binary logical values. Typical digital integrated circuits are used in computers, TV games, watches and so on.

While linear integrated circuits in the past have been the norm in most types of circuits, it is becoming more and more usual to see digital integrated circuits take their place in a digital representation of analogue functions. Perhaps the best example of digital integrated circuits encroaching on previously linear analogue functions is the case of the compact disc audio system, where high quality audio sound (a purely analogue parameter) is now undertaken totally by digital devices.

Operational amplifiers

Although used to perform accepted amplification functions in
the AC sense, operational amplifiers were originally developed
to perform mathematical operations on DC voltages. For this
reason they are typically used in equipment for measurement
and control of process variables in industrial and similar applications.

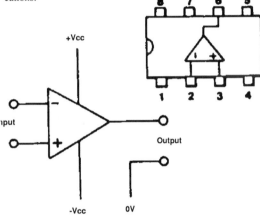

An operational amplifier is essentially a very high gain
amplifier around which negative feedback can be applied, the
gain characteristics depending on the nature of the components
in the feedback circuit.

Differential input amplifiers

Basic operational amplifier is, in fact, a differential input device.
Relationship between its output and input is given by:

$$Vo = A(e2 - e1)$$

where A is amplifier gain, Vo is output voltage, e1 is instantaneous voltage 1, and e2 is instantaneous voltage 2.

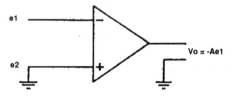

Generally the amplifier operates with its positive input
connected to the common rail, so e2 is zero, and:

$$Vo = -Ae1$$

the minus sign indicating phase inversion.

Inverting amplifier

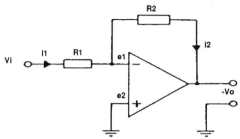

In the inverting amplifier, negative feedback is used. Output voltage Vo is always of opposite polarity to input voltage Vi. In the ideal condition amplifier gain is infinite which, as e2 is zero, tends to make e1 equal to e2. Thus i1 also equals i2. Therefore:

$$\frac{Vi}{R1} = \frac{-Vo}{R2}$$

and:

$$\frac{-Vo}{Vi} = \frac{R2}{R1}$$

As gain equals output divided by input (that is, Vo/Vi) gain can be seen to be dependent only upon the ratio of the feedback and input resistors (R2/R1). By suitable selection of these components amplifiers can be made to carry out mathematical operations including addition, subtraction, integration and differentiation.

Addition function

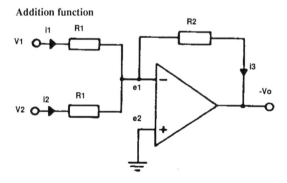

Assuming the ideal condition: e1 equals e2 equals zero, then i1 plus i2 equals i3 and, if the two resistors R1 are of equal value, then:

$$\frac{-Vo}{R2} = \frac{V1}{R1} + \frac{V2}{R2} = \frac{V1}{R1} + V2$$

Therefore:

$$-Vo = \frac{V1 + V2}{R1} \times R2$$

Subtraction function

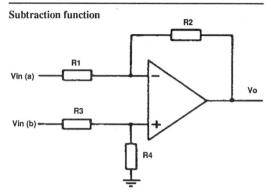

The subtractor uses the principles of the differential amplifier, incorporating functions of a 1:1 inverting amplifier and a 1:1 non-inverting amplifier, their output effectively superimposed on one another.

Inverting amplifier function is seen as being effectively the top half of the subtractor, providing an output which is the inverse of the input:

Vout = -Vin

Similarly, non-inverting function is the lower half of the subtractor, providing an output equal to its input:

Vout = Vin

Superimposing one on the other:

Vout = [Vin(a)] + [-Vin(b)]

Therefore:

Vout = Vin(a) - Vin(b)

The integrator

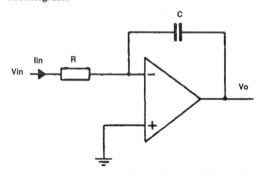

In this circuit a basic RC timing network is used to cause the output to change or ramp at a rate which depends on the value of input voltage. This is the basic action of the integral element of an electronic controller.

Circuit is a simple inverting amplifier with capacitive feedback. If resistor R and capacitor C are fixed then rate of change of output is dependent on input current Iin, the value of which is determined by input voltage Vin.

The differentiator

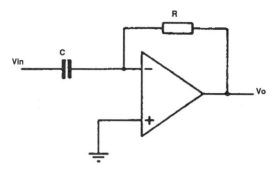

Gain of a basic inverting amplifier is dependent on ratio of feedback and input resistances but, as the input element of the differentiator is a capacitor (and thus is reactive not resistive), gain of the circuit is dependent on rate of change of voltage at the input. If the rate of change of input voltage doubles then output voltage value doubles. This is the basic action of the derivative element of an electronic controller.

Logic

Logic may be described as the art or science of reasoning. It is normally used at design stages, but can be used to improve reliability of performance and safe operation of control systems.

Industrial control systems

In practice the process to be controlled is often a complete plant, part of a plant, or a single machine. Desired states of the process are determined by setting switches, while actual states are detected by sensors. Desired and actual states are then compared in a logic unit and a control signal is derived which helps to reduce any difference from desired states.

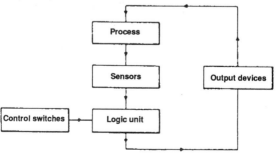

Pressure-operated switches and mechanically-operated limit switches are examples of sensors typically used. Logic unit output actuates output devices in the forms of electro-mechanical relays, solenoid-operated valves, or pneumatically-operated valves.

Switching techniques are involved. Any switch, at any given instant of time, is in one of two possible conditions: on or off. Thus the pattern of the logic unit must follow a binary digital pattern.

Binary system

Any system of counting possesses certain features:

● a symbol to represent zero
● a set of numbers
● a base
● place values.

In the binary system two symbols only are used, by convention: 0, representing zero, and 1. Any number may be represented by these two symbols.

Where decimal numbers have a base of 10, in which each column advancing to the left represents a ten-fold increase in place value, the binary system has a base of 2, in which each column advancing to the left represents a doubling of place value. Thus, in binary, the right-most column has a place value of 2, the next column to the left has a place value of 4, the next a place value of 8, next a value 16, next 32, and so on. So, a number such as 27 in decimal is represented by 11011 in binary:

$$(1 \times 16) + (1 \times 8) + (0 \times 4) + (1 \times 2) + (1 \times 1)$$

More columns, advancing to the left, are required for greater-valued numbers. For example the decimal number 1449 is represented by the binary number 10110101001. Such numbers are cumbersome and so are often transposed to octal numbering form for convenience. Here, eight becomes the system base, so 0 represents zero while number symbols are 1, 2, 3, 4, 5, 6 and 7. First step in transposition of a binary number to an octal number is to separate binary symbols into groups of three binary digits (bits). Thus 10110101001 becomes: 010 110 101 001. Note: a 0 has been added to complete the left-hand group. Decimal equivalents of the individual groups of three bits become the symbols now used, so 010 110 101 001 becomes: 2651 in octal.

Decimal/binary/octal conversion table

Decimal	Binary	Octal
0	0	0
1	1	1
2	10	2
3	11	3
4	100	4
5	101	5
6	110	6
7	111	7
8	1000	10

9	1001	11
10	1010	12
11	1011	13
12	1100	14
13	1101	15
14	1110	16
15	1111	17
16	10000	20
17	10001	21
18	10010	22
19	10011	23
20	10100	24
21	10101	25
22	10110	26
23	10111	27
24	11000	30
25	11001	31
26	11010	32
27	11011	33
28	11100	34
29	11101	35
30	11110	36
31	11111	37
32	100000	40
33	100001	41
34	100010	42
35	100011	43
36	100100	44
37	100101	45
38	100110	46
39	100111	47
40	101000	50
41	101001	51
42	101010	52
43	101011	53
44	101100	54
45	101101	55
46	101110	56
47	101111	57
48	110000	60
49	110001	61
50	110010	62
51	110011	63
52	110100	64
53	110101	65
54	110110	66
55	110111	67
56	111000	70
57	111001	71
58	111010	72
59	111011	73
60	111100	74
61	111101	75
62	111110	76
63	111111	77

Logic gate symbols

Logic gates form the fundamental electronic parts of logic circuits. Symbols used to represent logic gates are frequently those recommended in American military standard Mil-Std 806B, although, strictly, those of British standard BS3939 should be used in any UK-developed artwork.

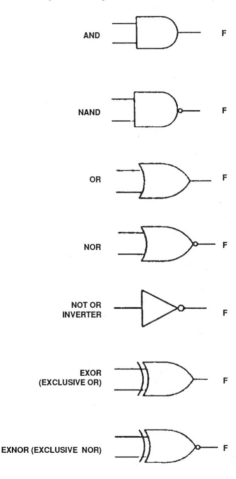

Logic modes

Logic systems normally comprise a number of logic elements or
gates, interconnected in such a manner that specified functions
may be performed. These gates are classified by the condition
of inputs in relationship to corresponding outputs. Various
gates with different characteristics are used. By convention it is
usual to specify binary 0 and 1 as:

Input signal applied	1
Input signal not applied	0
Output exists	1
Output does not exist	0

The AND function

Symbol

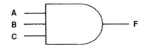

Equivalent circuit

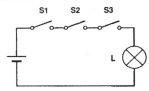

A circuit comprising switches in series with a lamp and cell will
perform the AND function. Lamp L lights if switches S1 AND
S2 AND S3 are closed. A true logic gate has inputs (A, B and
C, say) in place of switches, while an output is in place of a lamp.
The AND function is defined as: if A AND B AND C are at logic
1, then F equals logic 1; otherwise F equals logic 0. This is
written mathematically as:

$$A \cdot B \cdot C = F$$

Dots between input letters denote the AND function.

A truth table showing all possible conditions of inputs and
outputs can be compiled. From this an appropriate logic circuit
can be developed.

AND gate truth table

A	B	C	F
0	0	0	0
0	0	1	0
0	1	0	0
0	1	1	0
1	0	0	0
1	0	1	0
1	1	0	0
1	1	1	1

The OR function

Symbol

Equivalent circuit

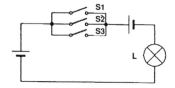

A circuit comprising switches in parallel performs the OR
function, in which lamp L lights if switch S1 OR S2 OR S3 is
closed. The OR function is defined as: if A OR B OR C is logic
1 then output F is logic 1; otherwise F is logic 0. This is
mathematically written:

$$A + B + C = F$$

Addition symbols indicate the OR function.

OR gate truth table

A	B	C	F
0	0	0	0
0	0	1	1
0	1	0	1
0	1	1	1
1	0	0	1
1	0	1	1
1	1	0	1
1	1	1	1

The NOT function

Symbol

Equivalent circuit

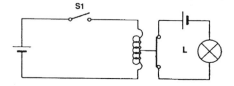

A NOT gate has only one input, and can be performed with a switch and an electromechanical relay which breaks the lamp circuit when switch S1 is closed. The NOT function is defined: if input A is logic 1 then output F is logic 0; if A is logic 0 then F is logic 1. Mathematically this is:

$$F = \overline{A}$$

where the bar over the A denotes the logical inverse of A.

NOT gate truth table

A	F
0	1
1	0

The NAND function

Symbol

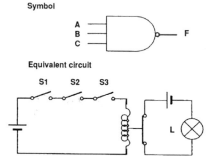

Equivalent circuit

The NAND function is the inverse of the AND function, which can be performed by series switches and an electromechanical relay to break the lamp circuit. The NAND function is defined as: if A AND B AND C are logic 1 then F is logic 0; otherwise F is logic 1. This is, mathematically:

$$\overline{A \cdot B \cdot C} = F$$

The bar indicates this is the inverse of the AND function: NAND = NOT AND.

NAND gate truth table

A	B	C	F
0	0	0	1
0	0	1	1
0	1	0	1
0	1	1	1
1	0	0	1
1	0	1	1
1	1	0	1
1	1	1	0

The NOR function

Symbol

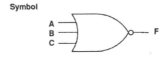

Equivalent circuit

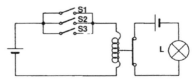

This is the inverse of the OR function, that is: NOR = NOT OR which can be performed by switches in parallel with an electro-mechanical relay to break a lamp circuit. When S1 OR S2 OR S3 is closed, the lamp is off. The NOR function is defined: if A OR B OR C is logic 1 then F is logic 0; otherwise F is logic 1. This is, mathematically:

$$\overline{A + B + C} = F$$

NOR gate truth table

A	B	C	F
0	0	0	1
0	0	1	0
0	1	0	0
0	1	1	0
1	0	0	0
1	0	1	0
1	1	0	0
1	1	1	0

The EXOR function

This function can be performed by two double-throw switches S1 and S2 in parallel. Lamp L lights only if switch S1 is closed and switch S2 is open, or if switch S1 is open and switch S2 is closed. If both switches are open, or both switches are closed,

Symbol

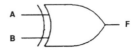

Equivalent circuit

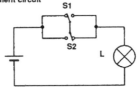

lamp L is off. It is sometimes known as the exclusive or function. The EXOR function is defined: if A and B are logic 1, or if A and B are logic 0, F is logic 0; otherwise F is logic 1. Mathematically this is:

$$A \oplus B = F$$

A circle around the addition sign denotes the exclusive or function.

EXOR gate truth table

A	B	F
0	0	0
0	1	1
1	0	1
1	1	0

The EXNOR function

Symbol

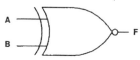

Equivalent circuit

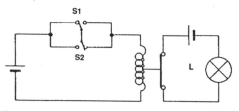

This is the logical inverse of the EXOR function and can be performed by two double-throw switches in parallel, along with an electromechanical relay in a lamp circuit. The EXNOR function is defined: if A and B are logic 1, or A and B are logic 0, then F is logic 1; otherwise F is logic 0. Mathematically, this is written as:

$$\overline{A + B} = F$$

EXNOR gate truth table

A	B	F
0	0	1
0	1	0
1	0	0
1	1	1

Hold or memory circuit

It is sometimes necessary to hold a circuit on after an initiating
signal is removed. A typical example is the hold-on device
associated with a simple starter for an electric motor, performed
with mechanical switches and an electromechanical relay in the
form of a coil and contacts. Pressing the start button energises
the coil, resulting in auxiliary contact AC closing. Even if the
start button is now released the coil remains energised through
the auxiliary contact. The coil also actuates a main contact
which supplies power to the motor. This memory is cancelled
once the stop button is pressed.

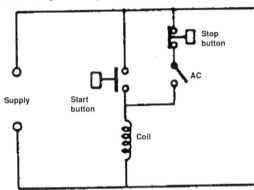

A memory circuit can be performed using logic gates, such
as two NOR gates. Momentarily pressing the set button applies
logic 1 to the first gate, output of which becomes 0. This logic
0 is applied directly to the second NOR gate, output of which
thus becomes 1. This output can be used to drive coils or relays,
and is also fed back to the other input of the first gate, holding
the circuit in this state.

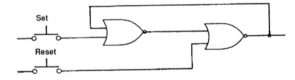

Memory is cancelled when the reset button is pressed,
because one input of the second gate becomes 1 so its output
becomes 0. As this is fed back to the first gate the circuit returns
to its original state.

Building gates

Two identical gates can be connected to perform a different
function. For example, two NOR gates in series perform the OR
function. Some logic systems are built solely from gates of just
one type, connected together to give all required logic functions.

Logic gates

Devices used as logic gates in logical process control circuits include:

● electromechanical relays

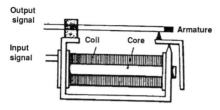

● diodes - valve or semiconductor

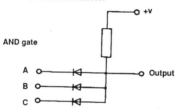

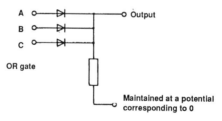

● transistors

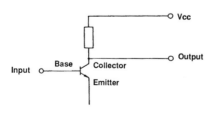

● turbulence amplifiers
● triode valves
● wall attachment amplifiers
● integrated circuit gates (made from transistors).

Microelectronic devices

Production of an integrated circuit microelectronic device involves a series of highly skilled and sometimes highly complicated processes. Cost of a device is between just a few pence and a few pounds, depending on complexity and manufacturing volumes.

A microelectronic device replaces a large circuit which, if made from discrete components, would be heavy, bulky, expensive and less reliable.

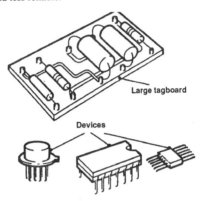

Large tagboard

Devices

Method of production

Two main steps are involved in manufacture of integrated circuit microelectronic devices. First is the manufacture of slices containing many hundreds of individual circuits:

● slices are cut from a silicon bar

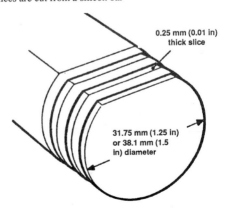

0.25 mm (0.01 in) thick slice

31.75 mm (1.25 in) or 38.1 mm (1.5 in) diameter

● epitaxial growth – slices are placed in a reaction chamber to undergo an epitaxial process in which one face of the slice is coated with silicon oxide

● photo-resist – slices are coated with a photo-resist

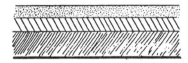

● alignment – each slice is aligned under a mask which contains a repeated pattern of part of the required circuit. Ultra-violet radiation is used to expose the pattern onto the slice

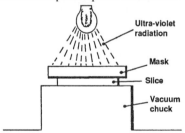

● developing and etching – exposed areas are developed and etched

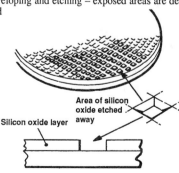

● diffusion – the slice undergoes a diffusion process.

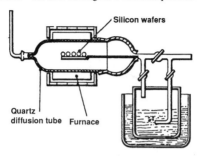

This has the effect of changing electrical properties of silicon slices at places where silicon oxide has been etched.

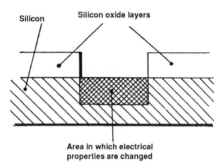

Following epitaxial growth, the series of processes: photo-resist, alignment, development, etch, and diffusion are repeated several times with different masks containing different parts of the required circuit, until a complete circuit comprising areas of varying electrical properties exists. Areas have properties of components such as resistors, capacitors, diodes and transistors.

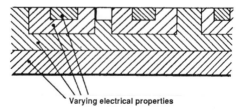

Quality
Photo-resist, alignment, development and etch processes are carried out in special dust-free areas. Special clothing is worn to protect slices from foreign bodies such as dust, hair, loose skin, grease.

Safety

Considerable use is made of chemicals during processes. Protective clothing is worn. Spilt chemicals must be cleaned up immediately, using specified materials. For splashes on skin or in eyes, wash well with cold water or rinse with eye douche and seek First Aid.

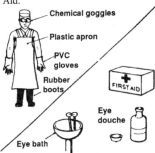

Furnaces used in epitaxial and diffusion processes operate at high temperatures. Do not touch the tube in the furnace in which slices are heated. Allow slices to cool after removal from furnace.

Second main step of manufacturing integrated circuit microelectronic devices comprises connection of leads and packaging:

● metallising – circuit components are linked up with a thin aluminium layer (0.001 mm) evaporated onto the slice. Circuit pattern is acid etched, then heated in a low temperature furnace to produce good electrical contact between silicon oxide and aluminium

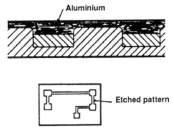

● scribing – lines are scribed between circuits

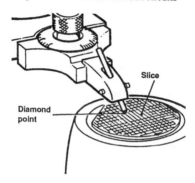

● dicing – each slice is broken up into its individual chips (also called dice). Each chip or die contains a complete circuit

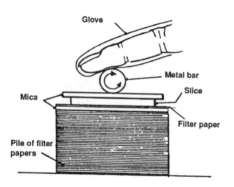

● alloying – each die is fitted into a package, by alloying it with the package material

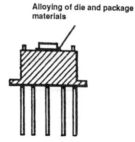

● die attach – a die is attached to the package by placing it in glass which is heated until soft, then cooled until set

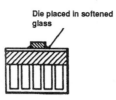

Die placed in softened glass

● bonding – fine leads (about 0.038 mm diameter) are bonded to the die and package terminations. Bonding is by thermo-compression or ultrasonic means

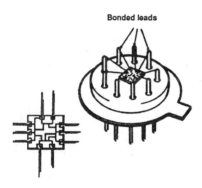

Bonded leads

● thermo-compression – leads are bonded by compression by a wedge under heat
● ultrasonic – lead and surface of the device are vibrated together at high frequency; resultant heat fuses the materials together

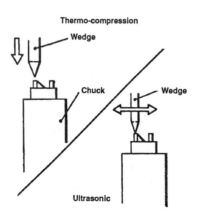

Thermo-compression

Wedge

Chuck

Wedge

Ultrasonic

● encapsulation – die and package are covered by a robust protective cover

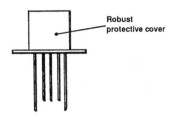

● test – devices receive tests to ensure performance and reliability.

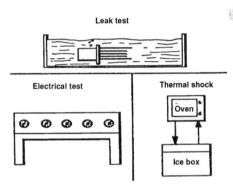

Thin-film modules

Thin-film modules form another means of miniaturising complex circuits into an integrated form. Unlike microelectronic devices, however, thin-film modules use discrete semiconductor devices. There are many manufacturing steps:

● substrate manufacture – substrate of a thin-film module is a plain piece of glass

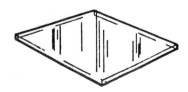

● evaporation – circuits are evaporated onto the glass in a vacuum chamber divided into three sections

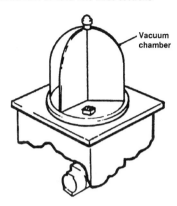

Vacuum chamber

● each section contains a mask on which three substrates are placed. A heating element below the chamber ensures correct temperature for processes in all three sections

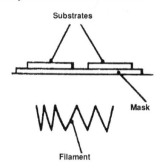

Substrates

Mask

Filament

● substrates enter the chamber in the first section and are not removed until all three processes, one in each section, are complete

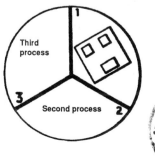

Third process

Second process

● evaporation – in section 1 of the chamber, resistors are evaporated onto the substrate

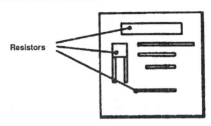

● in section 2, gold contacts are evaporated on

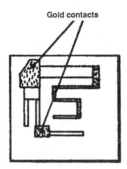

● in section 3, silicon monoxide is evaporated onto resistors. This protects them from damage and prevents later contamination

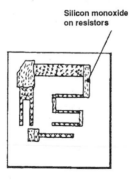

● during evaporation processes sections of the chamber are filled with the material evaporated on the substrate. This means the inside of the chamber is coated with the materials and has to be cleaned regularly

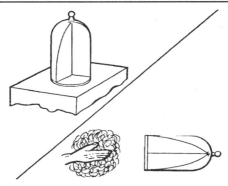

● module assembly – a substrate is assembled to a base with leads on a jig

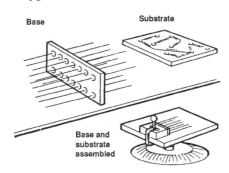

● leads are cut to length

● leads and discrete component leads are insulated and lead ends are prepared ready for soldering

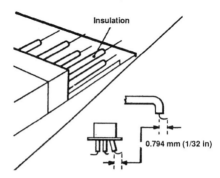

Insulation

0.794 mm (1/32 in)

● leads and components are hand soldered to gold contacts using a soldering iron bit at a high temperature (325° to 350° C)

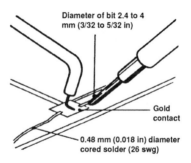

Diameter of bit 2.4 to 4 mm (3/32 to 5/32 in)

Gold contact

0.48 mm (0.018 in) diameter cored solder (26 swg)

Important
Soldering iron bit must not be held against the substrate for more than the specified time – or heat will damage the substrate. Soldering is performed in an ordered way, starting nearest the base and finishing furthest from it.

● the assembly is fitted into a can which protects the circuit from damage.

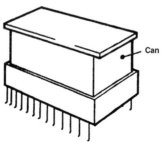

Can

Common wire and cables

Wire may be a single-strand of solid soft copper, or comprise a group of copper wire strands or threads.

Single-strand wire
Used where rigid wiring is necessary. It may be insulated with varnish or enamel paint. Always insulated when used to produce coils. Used uninsulated for the leads of resistors, capacitors and so on.

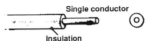

Single conductor

Insulation

Multi-strand wire
Used when flexibility is essential. The conductor is covered by rubber or plastic. For certain applications woven glass fibre or metal sheathing may encase the insulation. Wire sheathing is frequently called braiding or sheathing.

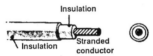

Insulation

Insulation   Stranded conductor

Both single-strand and multi-strand conductors are classified by the nominal cross-sectional area in mm2. Single-strand wire is classified to a standard wire gauge (SWG). Refer to SWG tables. Both single-strand and multi-strand wires may be given a code number representing cross-sectional area.

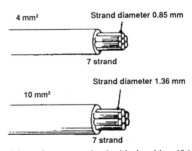

4 mm²   Strand diameter 0.85 mm

7 strand

10 mm²   Strand diameter 1.36 mm

7 strand

Flexible wiring often uses wire braid sheathing if it is required to conduct heavy current, or reduce electrical interference. Sheathing does not appreciably reduce flexibility.

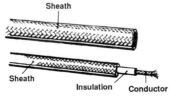

Sheath

Sheath   Insulation   Conductor

Types of cable

Two or more wire conductors, prefabricated together by a manufacturer, are called a cable. The same description applies when the process is accomplished by an assembler. Where wires are bundled together and tied (laced) by an assembler, resulting product is called a cableform, a harness, or a loom. A number of types is common:

● two-core flat – insulated and sheathed (also called flat twin)

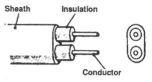

● three–core flat – as above but with an uninsulated earth continuity wire in same sheath

● parallel twin – cores easily separated without damage to insulation of either

● three-core – insulated, sheathed, unarmoured

● multi-core – insulated, sheathed conductors

● multi-core – insulated, flexible cable, braided sheath

● three-core mineral – insulated, copper-clad

● screened cable – insulated, multi-cored single conductors, braided copper wire sheath.

Coaxial cable

This is screened cable with a single central conductor of stranded
or single wire surrounded by concentric layers of insulation and
braided sheath. Braiding is used as a second conductor. For
radio frequency applications a special high quality central
insulation is used, which is much thicker. For very high
frequencies the outer conductor may be continuous copper tube.

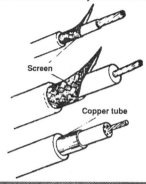

Screen

Copper tube

Current carrying capacity

This is the maximum permitted current a conductor can carry. It
is limited by the temperature which materials used can withstand. Greater current causes overheating of wire and insulation
and a fire could result.

In a three-core cable live and neutral conductors must each
be capable of carrying the normal load current. Earth conductors must be capable of carrying maximum fault current. A
conductor's current carrying capacity is less when the conductor
is part of a cable. When two or more cables are run within the
same conduit or trunking, current carrying capacity is reduced.
Ambient temperature around a cable also affects its current
carrying capacity – hotter the ambient temperature, lower the
capacity.

Wire tables

IEE wiring regulations contain tables showing current carrying
capacities of wires and cables.

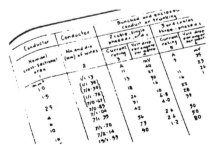

Wire identification

Wire may be identified in two general ways:

● numbering code – insulated wire may have a number printed at regular intervals in its insulation. Alternatively it may be identified by sleeving with pre-printed adhesive tape. Numbers correspond to those allocated on wiring diagrams or wiring routing schedules

● colour code – insulation of wire may be coloured on manufacture. Colour may be solid corresponding to a numerical code, or have coloured bands.

Multi-coloured insulations may follow a colour-code similar to resistor and capacitor colour-codes so, say, a red/green/blue coded wire is wire number 256, while a yellow/brown/green coloured wire is wire number 415).

Coloured marker sleeves may be fitted over each end of the wire, each colour conforming to standard code. For example wire number 15 has first a green marker sleeve then a brown marker sleeve. Colours are read from the outside in.

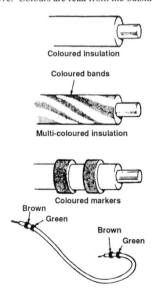

Coloured insulation

Coloured bands

Multi-coloured insulation

Coloured markers

Brown
Green
Brown
Green

Prefix letters are used to describe duties of the circuit formed by a wire. Such prefix letters are:

● H for AC supplies
● J for DC supplies
● N for tap change control, with numbers to identify individual wires.

Important
Beware of non-standard codes on internal wiring of imported equipment.

Ribbon cable
These cables comprise a number of thin, flat lengths of copper conductor, encased in plastic insulation. Such cables are thin and flexible, forming a simple method of creating cableforms and may be plain, or colour coded.

Removal of insulation
Cable insulation is removed so connection may be made to the conductor. Sufficient insulation should be removed from multi-cored cable to allow:

● easy access of individual wires to points of connection
● enough bare wire for the connection to be made securely.

Never remove more insulation than necessary.

Removing sheath
Sheath may be removed by:

● splitting the sheath along cable length

● peeling back the outer case and cut away unwanted insulation. Take care to avoid damage to wire insulation, particularly when making the cut along the cable.

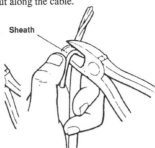

Sheath

Where there are several layers of insulation, each layer must be removed separately.

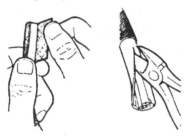

Removing varnish
Varnish may be removed by:

● burning with methylated spirits
● rubbing with emery paper

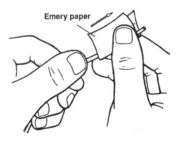

Emery paper

● scraping with a knife.

Important
Care must be taken with the latter two methods of removing varnish to avoid reducing diameter of conductor by removing copper along with the varnish.

Removing enamel paint
Enamel paint insulation may be removed with the aid of an
approved chemical stripper.

Important
When using this method care must be taken to eliminate all
residual chemical, otherwise corrosion to the conductor may
occur later.

Insulation stripping tools
There are three main types of tools used to strip soft insulations:

● precision strippers
● thermal strippers
● mechanical strippers.

Precision strippers
Jaws of this tool have V-shaped notches to cut insulation. An
adjusting screw operates as a stop to set closure of jaws to
accommodate a range of wire diameters. To strip insulation:

● turn the adjusting screw so the closed jaws cut the insulation
without penetrating the conductor

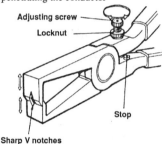

● tighten the lock nut on the adjusting screw
● place wire in the jaws

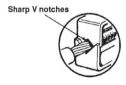

● squeeze the handles and rotate the strippers through a half-
turn
● pull the wire away from the tool to remove insulation.

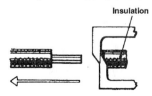

Alternative designs of precision stripper are available. In the tool shown, circular sharp edges within the jaws cut the insula-

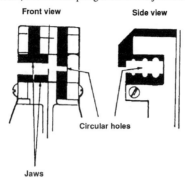

Front view   **Side view**

Circular holes

Jaws

tion automatically, without cutting the conductor.
    To strip insulation:

● squeeze the handles of the tool – the wire is thus gripped by the jaws

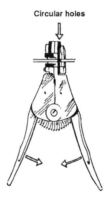

Circular holes

● squeeze   the handles further – the sharp edges cut the insulation

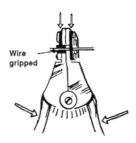

Wire
gripped

● release the handles, then turn wire through a quarter-turn
● squeeze the handles until the jaws separate, removing insulation from the wire.

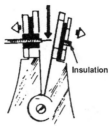

Stripped end

Insulation

Thermal strippers
This tool removes insulation by heating it with electrodes.

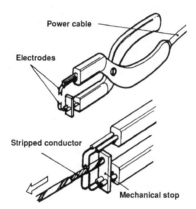

Power cable

Electrodes

Stripped conductor

Mechanical stop

Mechanical strippers
This tool is the least preferred of those in use.

Stripping insulation

A ragged edge to the cut insulation, nicked or broken strands are unacceptable.

Insulation should be cleanly cut and wire strands should be intact and undamaged.

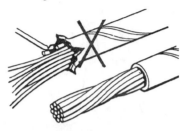

Forming pigtails in coaxial cable braiding

Using a plunging tool:

● remove outer insulation
● push braid back slightly to loosen it
● push the plunging tool under the braid until the tip eases between the braid weave
● press the plunger gently inward to ease the dielectric wire from the braid
● straighten out the braid and twist into a pigtail.

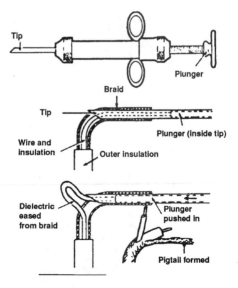

Using a screwdriver:

● remove outer insulation
● push the braid back towards the outer insulation until a small gap is formed in the braid
● bend the dielectric wire so a small loop protrudes through the gap made
● push the shaft of a small screwdriver under the loop and lever the dielectric wire through the gap
● straighten out the braid and twist into a pigtail.

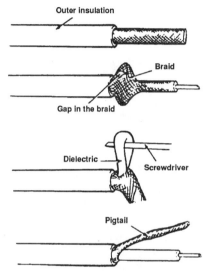

Using an awl:

● remove outer insulation
● comb strands of braid with the awl until they are of a sufficient length
● straighten strands and twist into a pigtail.

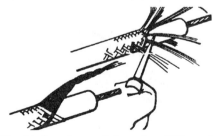

Pigtails, when formed, are ready for a crimped fitting or tinning with solder.

Coaxial cable bonding

Where many coaxial cables run parallel in a cable, it is sometimes useful to bond their screens through their pigtails. This is
a straightforward process:

● form braiding of each cable into a pigtail

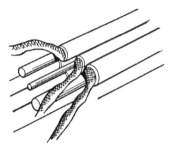

● bind together all pigtails, with a length of tinned copper solid
wire, not less than 2.5 cm from insulation

● solder connection, ensuring solder flows round each pigtail

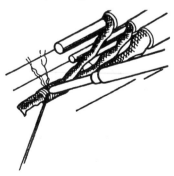

● sleeve joint and make it firm with the cableform by binding or sleeving. Heat-shrink sleeving is recommended.

Sleeving

Sleeves are used to insulate electrical joints thus reducing risk of short-circuits. Marker sleeves are used to identify wires. Sleeves may be used to bind wires into a cableform. Sleeving is often used to protect cableforms.

Fitting marker sleeves with a marker tool:

● push markers over the spike of the tool until they rest over the groove

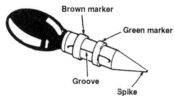

● push wire into the groove so markers are round the wire
● pull off markers and wire together.

Pull markers and wire

Fitting marker sleeves with a bullet tool:

● fit the bullet into the post

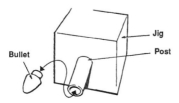

● push marker sleeves on to the post (sufficient sleeves for more than one wire may be positioned at one time)
● remove the bullet

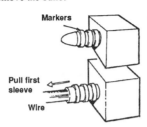

● push a wire into the hollow end of the post
● pull required sleeves onto the wire
● remove the wire.

Fitting marker sleeves with a sleeve stretcher.

A typical sleeve stretcher has three spikes which move apart as the handles are squeezed together.

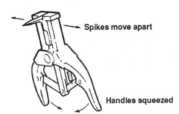

Method of operation is:

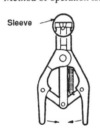

● place the sleeve over the spikes

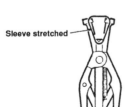

● squeeze the handles together to stretch the sleeves – do not over-stretch

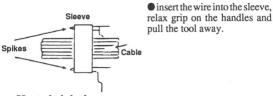

● insert the wire into the sleeve, relax grip on the handles and pull the tool away.

Heat-shrink sleeves
To fit this type of sleeving proceed as follows:

● obtain heat-shrinking sleeve of the diameter specified and cut to length
● slide sleeve into position
● heat the sleeve with a hot air gun, playing the heat evenly over the entire surface of the sleeve
● when the sleeve has shrunk into position, switch off hot air gun. Check the sleeve shrinks tightly into position as it cools.

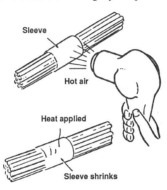

Safety
The hot air gun can be dangerous. Direct it only on the sleeve.

Wire terminations

Although many types of wire terminations exist, ground rules for all types are similar:

● cut back insulation to a suitable distance, taking care not to damage conductors. A conductor which is nicked with a knife, or grooved with a blunt cutting edge, is more likely to fracture in service than an undamaged one

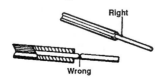

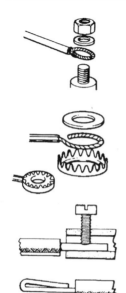

● for small cables, single or multi-strand conductors may be formed into a loop with rounded-nose pliers. Ensure direction of loop is such that tightening nut onto termination does not open the loop. Dipping loop in molten solder makes it stronger, but causes loss of flexibility, so failure may occur under vibration.

Use a special claw washer to get maximum connection. Lay looped conductor in washer, place plain washer on top and squeeze metal claws flat using correct tool – not ordinary pliers

● if pinch screws are used to hold wire, take care not to damage conductor by excessive tightening

● a single conductor should be doubled back on itself so the screw squeezes two strands and so is less likely to cause damage.

A stranded conductor should:

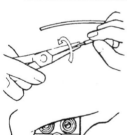

● be twisted to tighten strands in the direction of the existing twist (known as the lay of the wires)

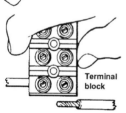

● be inserted as far as possible into a connector so the pinch screw does not clamp conductor end. If it does strands may untwist and connection may become loose.

Terminal block

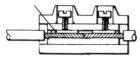

Clamp plate

Some types of terminal block have clamp plates which press against the conductor, so preventing damage by screw.

Wiring accessories

Tagged cable ends are used when wire is to be connected to a screw terminal. A suitable size must be chosen to fit cable and terminal. Most types are flat to allow panel fixing.

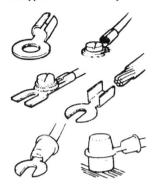

If frequent disconnection is likely, open-ended tags are best. Insulated types require clearance given by pillar terminals.

How tags are fitted to wire depends on tags used, although many may be soldered or crimped to conductors.

Crimped joints

Crimped joints are often used to connect wires to each other or to tags. Connectors used are called crimp fittings.

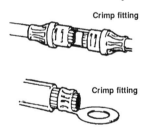

Crimp fitting

Crimp fitting

Wire is inserted into the fitting and the fitting is compressed (crimped) using a crimping tool.

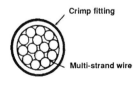

Crimp fitting

Multi-strand wire

A badly crimped joint cannot be re-worked. Scrap fitting and use new wire.

Crimping shapes are determined by a crimping tool's jaw shapes.

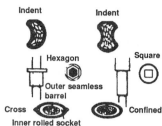

To make a crimped joint:

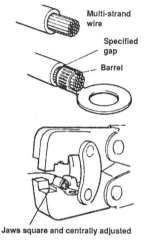

● strip insulation from wire end
● if multi-strand wire is used, twist conductors

● push wire into terminal, leaving specified gap between insulation and terminal barrel

● position tool jaws over the barrel, ensuring jaws are square to barrel and central

Jaws square and centrally adjusted

● make crimp by squeezing tool handles
● release tool by squeezing tool handles still further. Open tool handles and remove tool from joint.

Make crimp

Taper pins are used in miniaturised circuits. Both conductor and insulation are crimped.

Crimping

Hand-operated crimping tools

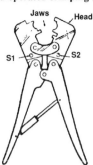

Jaws   Head

S1   S2

A typical tool has a crimping head which contains a set of jaws. Jaw shape determines crimp shape. Head and hence jaws of a tool may be removed by withdrawing screws S1 and S2. A head with different shaped jaws may then be secured to the tool.

Tool operates by squeezing handles together. Jaws then move together, grip, then crimp the fitting. Once a crimp is made the tool locks in position. Tool is released by squeezing handles still further.

Safety
Care must be taken when using this type of crimping tool not to trap a finger, as operation cycle of tool is non-reversible. Once handles are squeezed together, jaws can only be released by applying further pressure to handles.

Wire-wrapped connections

A technique of wrapping posts with wire is used to prototype circuits.

Several turns of single strand wire are wrapped, under pressure, round a metal post using a special wrapping tool. The metal post is called the wrapping post. A good conductive joint is formed by pressure of wrapped wire on corners of the post.

Wire
Wrapping post

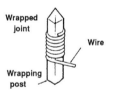

Wrapped joint

Wire

Wrapping post

Post corners dig into the wire and lock the joint. Wire is indented at contact points with post corners. Wrapping posts are usually square or rectangular in cross-section.

Wire is indented

Wire becomes indented

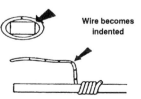

Wrapping tools

Wrapping tools can be electric or pneumatic. Operation of each is similar. An electric wrapping tool is described.

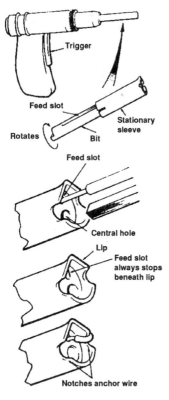

Wrapping wire is fed into a feed slot and the tool positioned so the wrapping post fits into the central hole.

The feed slot automatically comes to rest under the raised lip after each joint is made. This enables the feed slot mouth to be quickly located for the next joint.

Wrapping wire is anchored on one of the notches on the stationary sleeves. When the trigger is pressed, the bit rotates inside the sleeve and around the wrapping post, so wire is wrapped around the post.

Making a wrapped joint

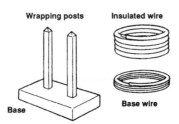

A length of single strand bare tinned copper wire and a length of single strand insulated wire are to be wrapped around two wrapping posts.

Tools required are:

Wire strippers   **Wrapping tool**

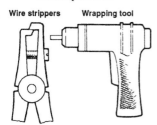

● wire stripper
● wire wrapping tool.

Bare tinned wire

Operation with bare tinned wire is:

● feed sufficient wire into the feed slot to give six turns on the wrapping post. Anchor the wire by placing it through the notch

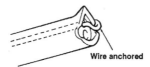

Wire anchored

● position the wrapping tool so the wrapping post is well up inside the central hole and the end of the bit is near the wrapping post base

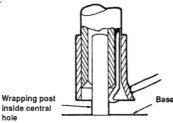

Wrapping post inside central hole   **Base**

● hold tool firmly and steadily, but do not bear down on it
● press trigger. As the joint is made, coiling action of the wire raises the tool clear of the joint.

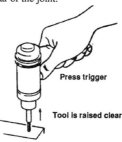

Press trigger

Tool is raised clear

Insulated wire

Operation with insulated wire is:

Wire and insulation into slot

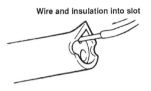

● strip enough insulation to give six turns of wire, using wire stripper
● feed wire and some insulation into slot
● feed enough insulation in, so when joint is made two post corners are in contact with insulation
● operate tool.

Multi-wrapped joints

Two or more joints can be made on the same wrapping post, provided the post is long enough.

Good and bad joints

A good joint is one in which:

● turns are spaced close together so no gaps are visible between
● there is no overlapping of bared wire

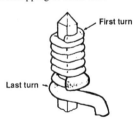

First turn

Last turn

● pull-off force (see later) is according to specification.

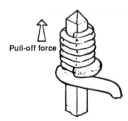

Pull-off force

A bad joint is one in which:

● bared wire overlaps

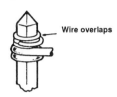

Wire overlaps

● turns are too widely spaced

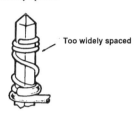

Too widely spaced

● pull-off force is below specification.

Never use a wrapping post with rounded edges. Never re-use wire unwrapped from a joint.

Pull-off tool

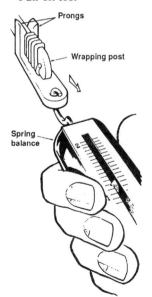

Prongs

Wrapping post

Spring balance

Use a pull-off tool to check binding strength of a wrapped joint. It has two prongs which fit either side of the wrapping post and under the joint.

A spring balance is attached to the other end of the tool. Pull the body of the spring balance so force indicated on the scale is equal to specified force.

Joints are tested on a sample basis. A joint must be able to withstand a force at least equal to the minimum specified.

Connectors

Connector is a general name for a plug or a socket.

A plug is a device to which a cable or individual wires can be attached. It has a contact pin or pins which fit into a socket to provide electrical connection. Contact pins are male contacts.

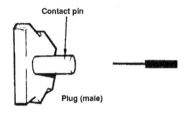

Plug (male)

A socket is a device with one or more female contacts, which mate with pin or pins of a plug.

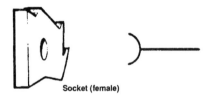

Socket (female)

Contacts are normally identified with small alphabetical letters printed on or formed in the body of the connector.

In general, connectors are of single-pole, multi-pole, or coaxial types.

Printed circuit board connectors

Connectors are usually cableform terminations.

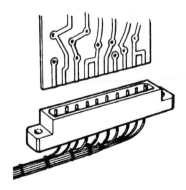

Ribbon cable connectors
These connectors are designed for an exclusive purpose and
cannot be used in any other application.

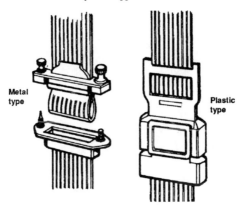

Metal type

Plastic type

Dual-in-line connectors
These connectors simplify component connection to printed
circuit boards. They accept dual-in-line packages, which may
be integrated circuits, relays, variable resistors, resistor networks and so on.

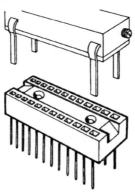

Networks of components may have certain leads removed
so they do not make contact with track on printed circuit boards.

Multi-pole connectors
A multi-pole is a plug or socket with a number of contacts.
Wires are joined to the connector contacts by crimping or
soldering. A type of multi-pole connector used for miniaturised
circuits is the high-density connector. These allow a large
number of low current circuits to be made in a small volume.

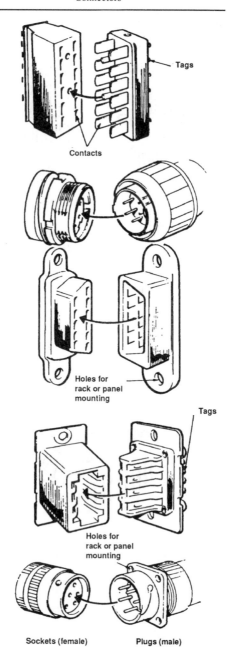

Tags

Contacts

Holes for
rack or panel
mounting

Tags

Holes for
rack or panel
mounting

Sockets (female)   Plugs (male)

Coaxial connectors

Coaxial connectors are plugs and sockets used to connect coaxial cables. Coaxial connectors may incorporate a means of mounting onto racks or panels. Typical fitting methods for coaxial plugs and sockets are:

● push fit

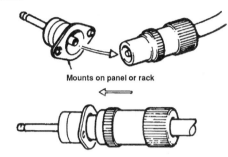

Mounts on panel or rack

● bayonet fit

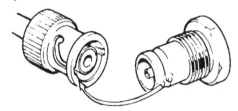

● screw fit

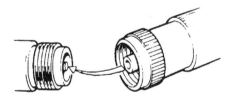

● spring fit.

Spring retainer

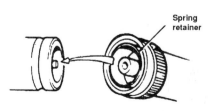

Polarisation of connectors

Purpose of polarisation is to ensure plugs and sockets can only be connected the correct way. Common types of polarisation are:

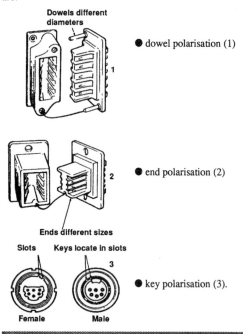

Dowels different diameters

● dowel polarisation (1)

● end polarisation (2)

Ends different sizes

Slots   Keys locate in slots

Female   Male

● key polarisation (3).

Multi-pole connector assembly

Essential tools required are:

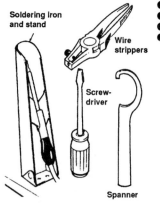

Soldering iron and stand

Wire strippers

Screw-driver

Spanner

● soldering iron and stand
● wire strippers
● screwdriver
● spanner.

The socket mounts on a rack or panel. The plug is fitted with a clamp to restrain cable movement.

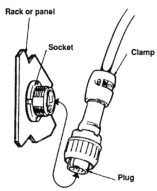

To assemble the plug:

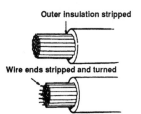

Outer insulation stripped

● strip away outer insulation from the end of the multi-core cable to the length specified

Wire ends stripped and turned

● strip and tin wire ends

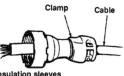

Clamp   Cable

Insulation sleeves

● slide clamp over cable

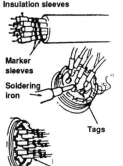

Marker sleeves

Soldering iron

Tags

Insulation sleeves

● place marker sleeves and insulation sleeves over wires as specified

● solder each wire of the multi-core cable to each tag in turn. This can be done either by reference to a wiring diagram or to a model

● push sleeves over tags

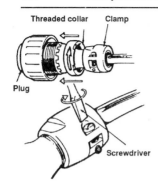

● locate clamp onto cable using screwdriver.

To assemble the socket:

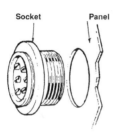

● remove nut and washer, then place socket through rack or panel

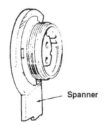

● place washer over socket and tighten nut using spanner

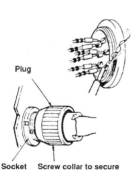

● wire the socket in the manner described for plug

● mate the two connectors by locating polarising plug keys with socket slots and screwing the collar over the socket to secure.

Electrical inspection

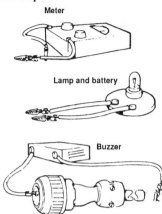

Use a buzzer, lamp or meter to check each connection.

High-density connector assembly

As area around each conductor in a high-density connector is too small to carry pin identification a datum mark is provided. Identity of each pin must then be ascertained by reference to a standard configuration.

Wires are fixed to pins by crimping. Special tools are required to insert and extract male and female pins.

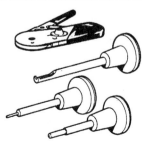

Order of assembly is:

● insert bared wire in pin end. Place both in crimping tool and close tool handles. Test crimp strength before proceeding.

● lay pin with lead attached in insertion tool. Pin shoulder should be outside tool face

● locate correct hole in body connector and insert pin. Press with insertion tool until pin clicks into position. Check pin is firmly home before proceeding to following pins

● fasten cable clamp to relieve tension between wires and pins

● fit plastic sleeve.

Note if other ends of wires are already terminated, the plastic sleeve must be fitted over wires prior to connector assembly.

Remove a pin by inserting extraction tool into connector hole. Push against sprung locating piece until pin slides out.

Covers

Covers may be used with connectors to prevent damage caused by movement of wires.

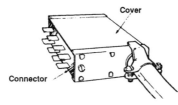

A typical cover is mounted as follows:

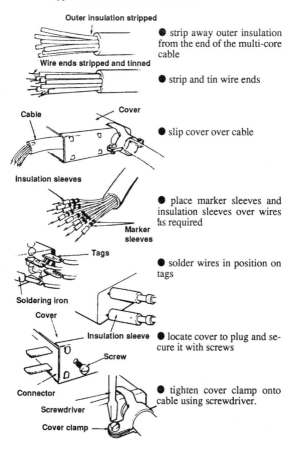

- strip away outer insulation from the end of the multi-core cable

- strip and tin wire ends

- slip cover over cable

- place marker sleeves and insulation sleeves over wires as required

- solder wires in position on tags

- locate cover to plug and secure it with screws

- tighten cover clamp onto cable using screwdriver.

Latches

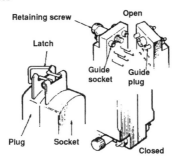

Latches may be used to lock the plug and socket together.

Coaxial connector assembly

Tools required for assembly of coaxial connectors are:

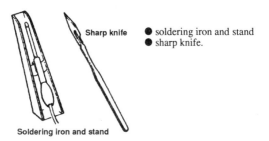

● soldering iron and stand
● sharp knife.

Assembly depends largely on connector type. Two types are described.

Push fit type with a rack or panel mounted socket
To assemble the plug:

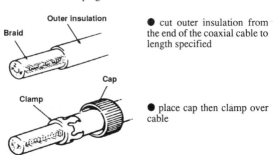

● cut outer insulation from the end of the coaxial cable to length specified

● place cap then clamp over cable

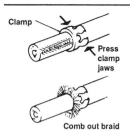

Clamp

Press clamp jaws

● hold end of clamp flush with end of insulation and press clamp claws lightly so they clasp insulation

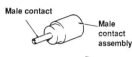

Comb out braid

● comb out braid and fold it over the top of the clamp

Male contact

Male contact assembly

● cut dielectric so the bared centre conductor is the same length as the male contact

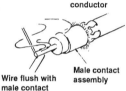

Tin centre conductor

● tin the centre conductor taking care not to damage insulation by excessive heating

Wire flush with male contact

Male contact assembly

● place male contact assembly over centre conductor and solder the end taking care not to leave iron in position too long

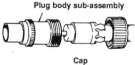

● trim excess braid

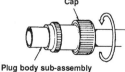

Plug body sub-assembly

● slide plug body sub-assembly over male contact assembly

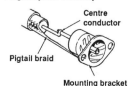

Cap

Plug body sub-assembly

● screw cap on to plug body sub-assembly.

Centre conductor

Pigtail braid

Mounting bracket

The socket has a mounting bracket attached to its body. Solder braid to bracket and centre conductor to female contact.

Spring fit coaxial type
To assemble:

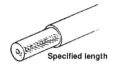

Specified length

● cut outer insulation from the end of coaxial cable

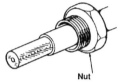

Nut

● slide nut over insulation

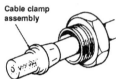

Cable clamp assembly

● place cable clamp assembly over braid and push it back against the insulation

Fold braid over

● comb out braid and fold it evenly over cable clamp assembly

● cut back dielectric to specified length, and tin centre conductor

Washer

● slide washer over dielectric so it is hard up against braid

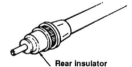

Rear insulator

● slide rear insulator over centre conductor so the insulator end butts flush against washer.

If the connector is a plug:

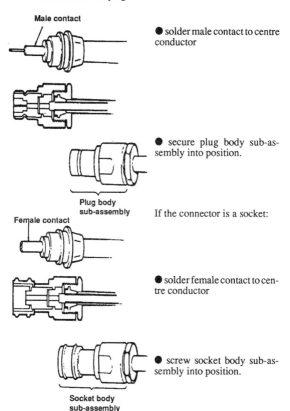

Male contact

● solder male contact to centre conductor

● secure plug body sub-assembly into position.

Plug body sub-assembly

If the connector is a socket:

Female contact

● solder female contact to centre conductor

● screw socket body sub-assembly into position.

Socket body sub-assembly

Cable glands

Glands are often used to protect a cable from mechanical damage at point of entry into equipment. The gland should securely retain outer sheath or armour of cable and maintain earth continuity between sheath or armour and equipment.

There are many types of gland units and reference should be made to manufacturer's literature when making or breaking these connectors.

Printed circuit boards

A printed circuit board (PCB) comprises a pattern of thin
metallic foil conductor bonded to a board of insulating material.
The metallic foil makes interconnections between discrete
components to form an electronic circuit.

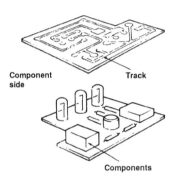

Component
side   Track

Components

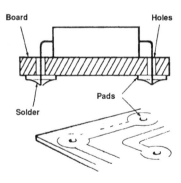

Board   Holes

Pads

Solder

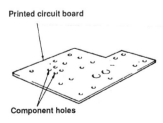

Printed circuit board

Component holes

The pattern is called track, while component connecting points are called pads, or lands. Components are normally mounted on the plain side of the board, although some printed circuit boards (called double-sided boards) have a copper track on each side. Some components are mounted in sockets or connectors.

All other components are mounted on terminal posts or directly to the board. Leads of board-mounted components are inserted in holes through the board and are soldered to pads usually after clinching or bending. Where components are mounted across conductive tracks care must be taken to follow instructions on their insulation.

A completely assembled PCB often constitutes a module.

Printed circuit board manufacture

Copper

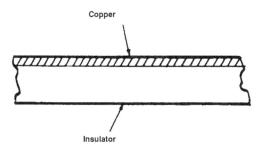

Insulator

Material used to make printed circuit boards is termed a laminate or copper-clad board. It comprises an insulator board with one or both surfaces completely covered in copper foil bonded to the surface. Insulator boards are usually made from one of two materials:

● resin-bonded paper
● epoxy glass fibre.

Pattern of the required track is initially drawn on paper, in black ink or tape, to form the master drawing. This type of drawing is often called artwork.

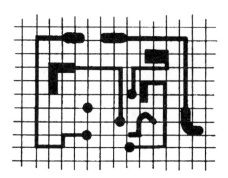

To make a printed circuit board, copper foil is removed from the laminate leaving only the required pattern of copper foil track. An etching process is used to do this.

To prevent the etchant material removing required track, areas of laminate forming track are first covered with a special etch-resistant chemical, commonly called resist.

Two main methods are used to manufacture printed circuit boards — the nylon screen method, and the photographic method.

Nylon screen PCB manufacture

This involves use of a porous screen of nylon stretched across the underside of a rectangular wood frame. Process involves a number of stages:

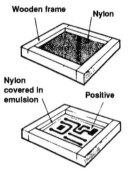

Wooden frame **Nylon**

Nylon covered in emulsion **Positive**

● screen is covered with a photographic emulsion and the photographic positive of the master drawing is placed on the screen

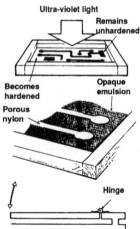

Ultra-violet light

Remains unhardened

Becomes hardened **Opaque emulsion**

Porous nylon

Hinge

● screen is exposed to ultra-violet light which hardens exposed emulsion — that is, emulsion round the track image is hardened, whereas emulsion under the track image remains unhardened

● developer is applied to wash away unhardened emulsion leaving hardened emulsion, in the form of a photographic negative of track, in position

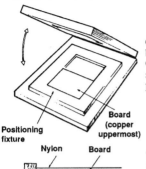

● wooden frame is hinged to a block of wood
● laminate is positioned under screen, with copper foil uppermost

Positioning fixture
Board (copper uppermost)

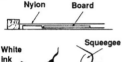

Nylon **Board**

● wooden frame is lowered so screen is almost in contact with laminate

White ink **Squeegee**

● etch-resistant ink is applied and is squeezed across the screen

Pattern formed on copper

● ink is squeezed onto copper foil, through the track pattern.

Once dry, board is ready for etching.

Photographic PCB manufacture

This involves coating laminate directly with a photographic emulsion. Process involves a number of stages:

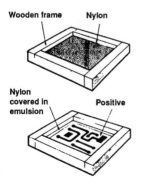

Wooden frame **Nylon**

Nylon covered in emulsion **Positive**

● laminate is sprayed with, or dipped in, light-sensitive photographic emulsion resist. This may be negative-resist (in which case following stage uses a photographic negative of the master drawing) or positive-resist (in which case a photographic positive of the master drawing is used). We shall assume negative-resist is used to coat the laminate
● negative of the master drawing is placed in close contact with laminate

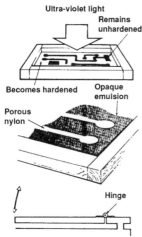

● negative and board are exposed to ultra-violet light which hardens exposed resist (if positive-resist is used, light softens exposed resist), thus resist along the printed circuit track is hard, while the rest is soft

● laminate is placed in a developer solution which dissolves away soft resist. A dye may be included at this stage, which dyes the printed circuit track. Laminate is now ready for etching.

Etching printed circuit board

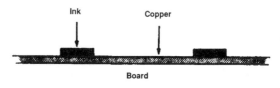

● board is sprayed with, or dipped into, a chemical which etches away bare copper while leaving copper under the resist in place
● board is thoroughly cleaned and resist is removed leaving the printed circuit pattern as copper on the board.

Edge connectors

When printed circuit boards are required for assembly into racks
or cabinets, they are usually manufactured with edge connectors
which mate with sockets. These edge connectors must be
durable and allow good electrical contact, so are normally plated
with gold or palladium.

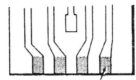

Edge connectors (gold or palladium)

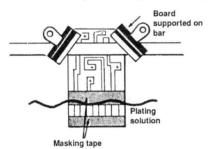

Board
supported on
bar

Plating
solution

Masking tape

Board is masked with tape so only parts required for edge
connectors are plated. It is placed in a bath and electroplated.
After plating masking tape is removed.

Roller soldering

After etching board is often roller soldered to tin track. Any edge
connectors are covered with masking tape and board is dipped
in flux.

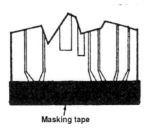

Masking tape

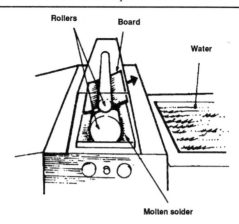

Board is fed track downwards between two rollers. Lower roller revolves in a bath of molten solder, so track becomes covered in an even layer of solder. Board is washed, cut to size and holes for components drilled.

External connections to printed circuit boards

Connections to printed circuit boards may be made by either pins or edge connectors. Typically. a rack-mounted printed circuit board is attached to a bracket which serves as a handle.

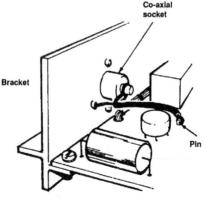

Printed circuit board

Edge connectors mate with a rack- or panel-mounted socket.

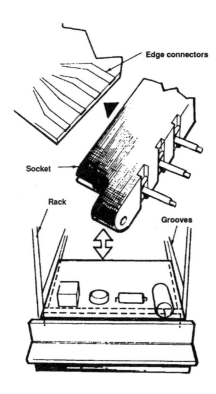

Printed circuit assemblies slide in grooves in a rack and can be fitted or removed quickly. Handles are often provided to enable printed circuit boards to be removed without too much handling.

Generally, edge connectors for boards are designed to fit particular racks or modules. These racks and modules are usually of standard types, defined in national and international standards documents.

Often edge connectors are loosely classified according to actual connector widths and four main widths are common:

- 0.156 inch
- 0.15 inch
- 0.125 inch
- 0.1 inch.

Apart from connector widths, standards usually define other parameters such as current rating, connector resistance and allowable temperature range.

Plated-through hole printed circuit boards

Printed circuit boards containing plated-through holes (PTH) are similar in appearance to normal double-sided boards. Plated-through hole printed circuit boards, however, have a continuous layer of metal lining barrels of holes.

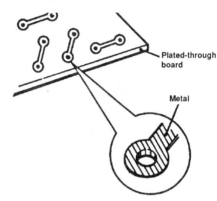

These provide electrical connections between printed circuit tracks on both sides of the board.

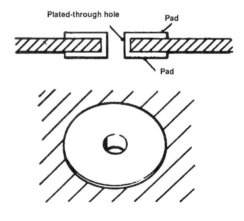

Enlarged view

They also may provide electrical connections between internal layers of multi-layered printed circuit boards.

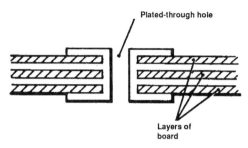

Plated-through hole

Layers of board

Multi-layered board

Production

Laminate comprises a sheet of epoxy glass fibre board, or other suitable material, clad on both sides with thin copper foil.

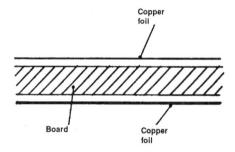

Copper foil

Board   **Copper foil**

Processes involve a number of stages:

● setting a datum — board is cut oversize. Holes are cut into it for use as a datum. These datum holes are used throughout manufacture to ensure all pads, tracks, holes and edge connectors are correctly aligned

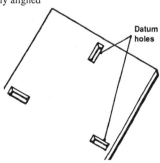

Datum holes

● grid system — more holes are drilled through the board where specified. These holes conform to a 2.5 mm (0.1 in) grid system. On subsequent processes this grid system is used to ensure pads are correctly positioned over holes. Holes are drilled on the intersection of two lines of the grid. Components leads are subsequently formed to grid sizes, to fit accurately to specified pad holes

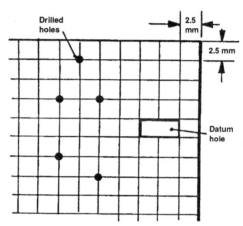

● cleaning — this process ensures no impurities remain on the board

Cleaning process

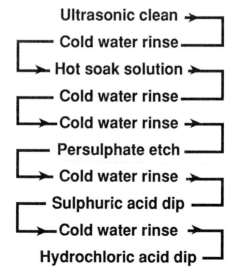

● exposure — board and mask are exposed to ultra-violet light, which hardens all resist except those masked

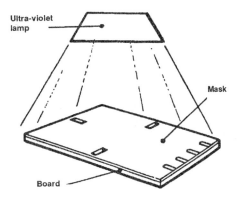

● developing — board is immersed in a developer which dissolves unexposed resist

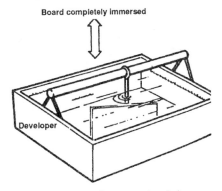

● after developing uncovered areas require plating

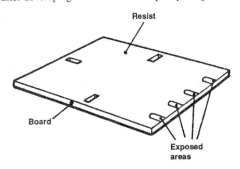

Safety
Protective clothing must be worn at all times. Boards must be handled carefully to avoid splashing harmful chemicals.

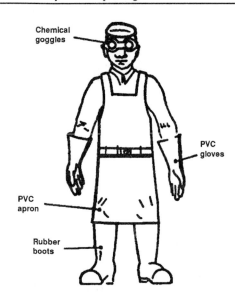

Once cleaned the board must be metallized immediately before boards are re-contaminated — re-contamination occurs even if boards remain clean simply because copper oxide forms on bare copper:

● metallising — stage one. Board is immersed in a catalyst solution in which minute particles of palladium are suspended. These particles become attached to all exposed board surfaces. Board is then rinsed thoroughly twice in cold running water

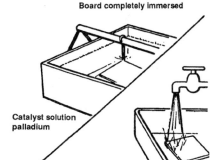

● accelerator — board is immersed in an accelerator solution for a specified period. This leaves the palladium coating ready to accept next stage

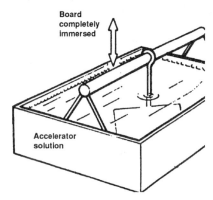

● metallising — stage two. Board is immersed in an electroless copper plating solution. Copper is deposited over the palladium to a thickness between 1 μm and 5 μm

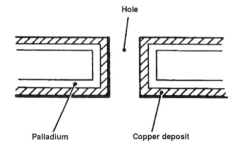

● electroplating — standard electroplating techniques are used to build up the copper layer to a thickness of around 0.025 mm

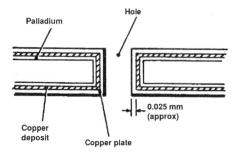

● resist — board is covered on both sides with a light-se. resist layer. Various methods are used, one of which is by r depositing a soft even layer onto one side of the board

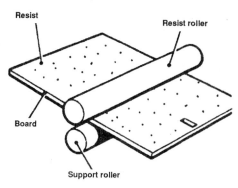

● this is allowed to dry and the operation is repeated for the other side

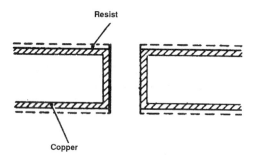

● masking — a photographic mask of gold contacts required is placed over the board

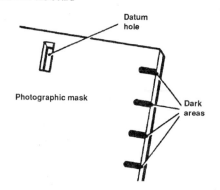

● gold plating — gold is electroplated onto exposed areas of the board

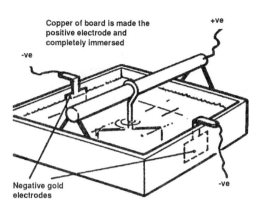

Copper of board is made the positive electrode and completely immersed

+ve

-ve

-ve

Negative gold electrodes

● all resist is chemically stripped from both sides of the board

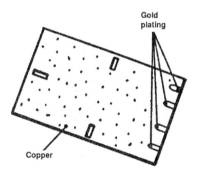

Gold plating

Copper

● gold areas are masked with protective tape and fresh resist is applied to both board sides as before

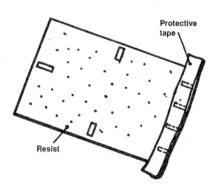

Protective tape

Resist

● tracks and pads — a photographic positive master of the track pattern is laid over the board

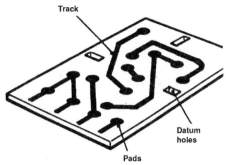

Track

Datum holes

Pads

● exposure — board and master are exposed to ultra-violet light

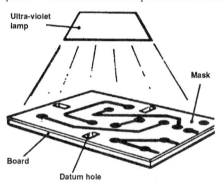

Ultra-violet lamp

Mask

Board

Datum hole

● developing — unhardened resist is dissolved away with a developer

Board completely immersed

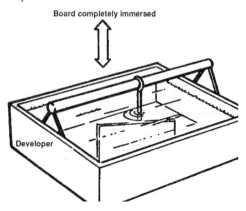

Developer

● board now has bare areas of copper which are to make up the track, surrounded by resist — on both board sides

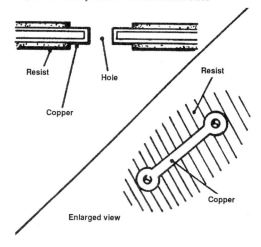

Resist

Hole

Resist

Copper

Copper

Enlarged view

● tin/lead plating — tin/lead alloy is electroplated onto bare copper on both sides of the board, including barrels of holes

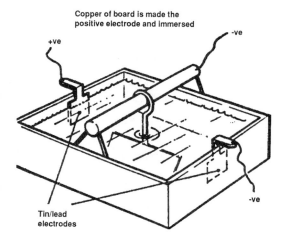

Copper of board is made the positive electrode and immersed

+ve

-ve

-ve

Tin/lead electrodes

● stripping — all resist is chemically stripped from both board sides. This leaves areas of copper coated with tin/lead or gold plating

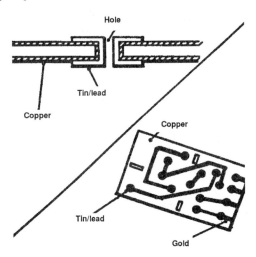

● etching — boards are sprayed with, or dipped in, an etchant to remove unwanted copper. Etchant removes bare copper but not gold or tin/lead plating, or copper underneath plating

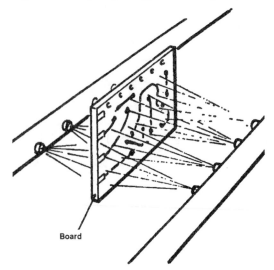

● cleaning — boards are thoroughly cleaned

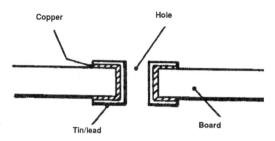

● legending — legend markings, denoting component positions and so on, are printed onto the board

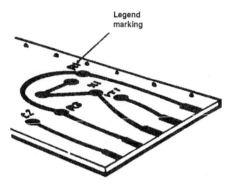

● trimming — boards are trimmed to specified dimensions. This removes datum holes which are no longer required.

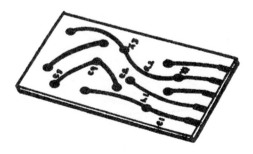

Multi-layered printed circuit boards

Multi-layered printed circuit boards comprise alternate layers of
copper foil and insulating material. There may be many layers
(up to 30) in one board and they are very closely packed. Total
thickness of board is usually no more than 2.4 mm.

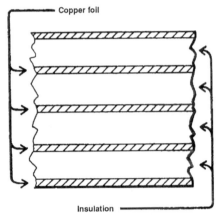

In the same way copper foil is etched to leave track and pads
on an ordinary printed circuit board, each layer of a multi-
layered printed circuit board is etched prior to bonding together
into a finished board. Printed circuit tracks are thus sandwiched
together between insulating layers within the multi-layered
board.

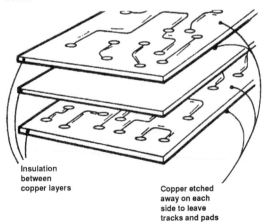

Advantages

Multi-layered printed circuit boards have many advantages, among the most significant are:

● replacement of complex inter-unit wiring, to make equipment lighter, smaller and cheaper

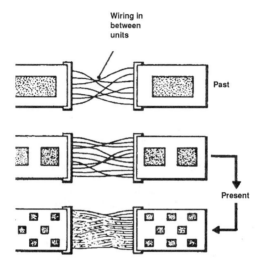

● compactness of on-board connections

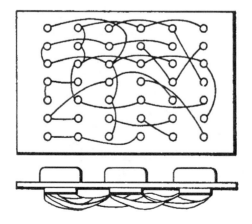

Same three modules connected with layered tracks of multi-layered printed circuit board

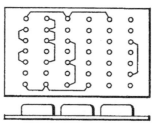

● as a result of shorter wiring, time for electrical pulses to travel from one component to another can be reduced. This is a very important factor in many types of equipment, such as computers, where speed of information must be as fast as possible

Computers used for fast data processing

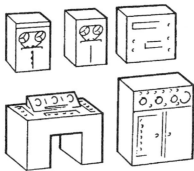

● internal layers may be left as complete or near-complete layers of copper foil. These may be used for earthing or screening purposes.

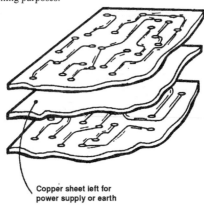

Copper sheet left for power supply or earth

Multi-layered board manufacture

Extremely complex and miniaturised circuits may be constructed using multi-layered boards. They are, however, complex in manufacture. Essentially, a multi-layered printed circuit board is made by joining together a number of thin epoxy glass fibre resin boards, or boards of a similar material.

Each board is covered on both sides with thin copper foil, and is known as a cured double-copper-clad board.

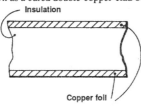

Insulation

Copper foil

Following processes are involved in manufacture of multi-layered printed circuit boards:

● stage 1 — accurate datum holes are made in each double-copper-clad boards. These ensure boards may be correctly aligned and positioned when joined together

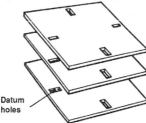

Datum holes

● stage 2 — each copper foil eventually forming an inner layer is separately printed and etched with its own track pattern. Method for this stage is same as for manufacture of ordinary printed circuit board. Note copper forming two outer layers of the whole multi-layered board is not printed or etched at this point

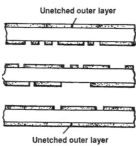

Unetched outer layer

Unetched outer layer

● stage 3 — plain sheets (that is, unclad and uncured) of epoxy glass fibre are drilled according to datum holes. These sheets are impregnated with epoxy resin and are known as prepreg sheets

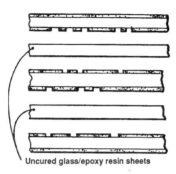

Uncured glass/epoxy resin sheets

● stage 4 — layers are fitted in order into a fixture comprising metal plates with accurately positioned alignment pins

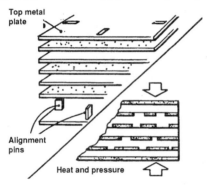

Top metal plate

Alignment pins

Heat and pressure

This sandwich of layers is placed in a press and heated under pressure. Prepreg sheets soften and run under pressure to fill air spaces between layers. At 165° C the prepreg cures, becomes hard and bonds layers together

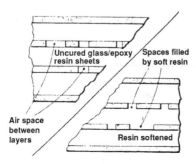

Uncured glass/epoxy resin sheets

Spaces filled by soft resin

Air space between layers

Resin softened

● stage 5 — board is removed from press and fitted to a jig. Holes are drilled where required

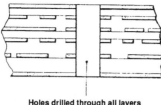

Holes drilled through all layers

● stage 6 — drilled holes are plated with copper by the plated-through hole method

Plating on surfaces of hole

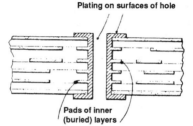

Pads of inner (buried) layers

● stage 7 — top and bottom board surfaces are printed and etched in the same manner as double-sided plated-through hole boards

Outer layer processed

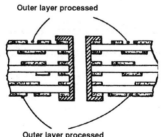

Outer layer processed

● stage 8 — legends are printed onto the board's surfaces where required.

Legend markings

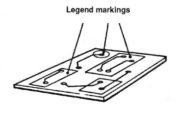

Number of layers

Multi-layered printed circuit boards are often referred to by the number of copper track layers incorporated.   Manufacturing method described produces a multi-layered board with an even number of copper layers.

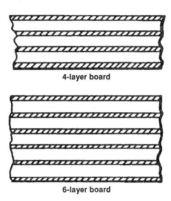

4-layer board

6-layer board

To produce a multi-layered printed circuit board with an odd number of layers one of the layers must be single-clad, with copper foil on one side only.

Test and inspection of printed circuit boards

To minimise risk of failure every sheet of copper-clad board is carefully inspected and tested before it is used for manufacture of printed circuit board.   This is especially important prior to manufacture of multi-layered printed circuit board.

● resist — board is covered on both sides with a light-sensitive resist layer. Various methods are used, one of which is by roller, depositing a soft even layer onto one side of the board

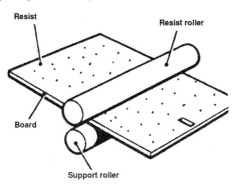

Resist

Resist roller

Board

Support roller

● this is allowed to dry and the operation is repeated for the other side

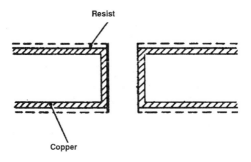

Resist

Copper

● masking — a photographic mask of gold contacts required is placed over the board

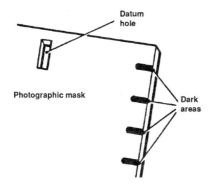

Datum hole

Photographic mask

Dark areas

● exposure — board and mask are exposed to ultra-violet light, which hardens all resist except those masked

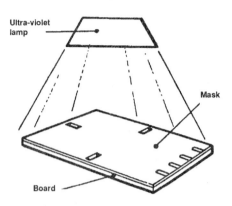

● developing — board is immersed in a developer which dissolves unexposed resist

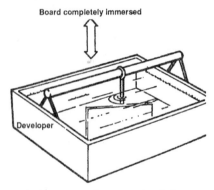

● after developing uncovered areas require plating

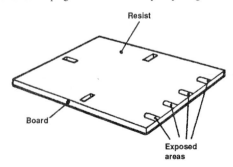

Safety
Protective clothing must be worn at all times. Boards must be
handled carefully to avoid splashing harmful chemicals.

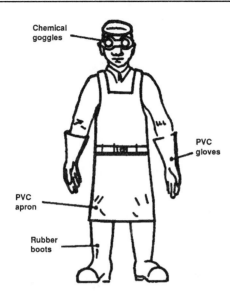

Once cleaned the board must be metallized immediately before
boards are re-contaminated — re-contamination occurs even if
boards remain clean simply because copper oxide forms on bare
copper:

● metallising — stage one. Board is immersed in a catalyst
solution in which minute particles of palladium are suspended.
These particles become attached to all exposed board surfaces.
Board is then rinsed thoroughly twice in cold running water

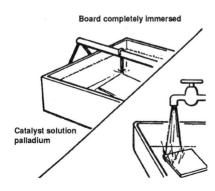

● accelerator — board is immersed in an accelerator solution for a specified period. This leaves the palladium coating ready to accept next stage

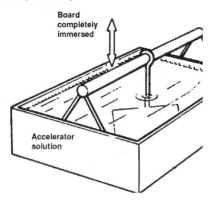

Board completely immersed

Accelerator solution

● metallising — stage two. Board is immersed in an electroless copper plating solution. Copper is deposited over the palladium to a thickness between 1 μm and 5 μm

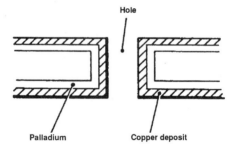

Hole

Palladium   Copper deposit

● electroplating — standard electroplating techniques are used to build up the copper layer to a thickness of around 0.025 mm

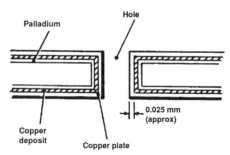

Palladium   Hole

Copper deposit   Copper plate

0.025 mm (approx)

Two properties of most importance are thickness and adhesion of copper foil. Sheets are also inspected for surface flaws.

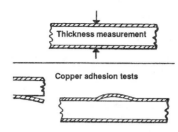

Sheets of prepreg are tested for curing time, resin content, resin flow and cleanliness.

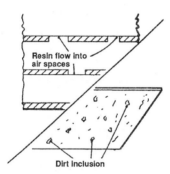

Completed printed circuit boards are examined for a number of things, including:

- broken tracks
- pinholes — no pinholes are allowed in edge connectors
- patches on track where there is no solder
- track lifting
- blistering of edge connectors
- hole positions and hole diameters.

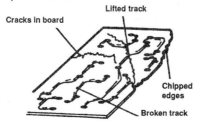

Inspection of multi-layered printed circuit board

Multi-layered printed circuit boards undergo up to three methods of test — visual, using X-ray equipment, electrical.

Visual inspection

Board is inspected for faults which may occur on ordinary printed circuit board, plus:

● microscopic examination of cross-sections of off-cuts to ensure bonding has cured properly

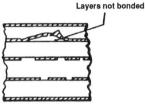

Layers not bonded

● checking for warping

● checking for corrugation.

X-ray inspection

Using special X-ray test and inspection equipment, multi-layered printed circuit boards are checked for:

● alignment of internal pads in relation to holes

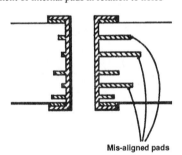

Mis-aligned pads

● internal wiring errors or failures

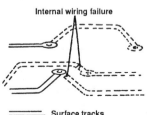

Internal wiring failure

Surface tracks
Internal tracks as shown by X-ray

● dirt or air spaces between layers

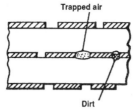

Trapped air

Dirt

● butt-joint failures, often caused by overheating during manufacture, component assembly or repair. In this type of fault internal pads break away from copper plating on hole barrels.

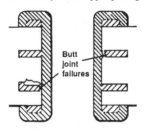

Butt joint failures

Electrical tests
Using appropriate test equipment, circuit continuity and breakdown of insulating materials can be checked.

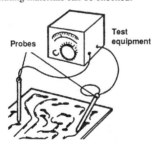

Test equipment

Probes

Quality
Only faulty multi-layered boards which are marginally warped or corrugated, and possibly those with wiring faults on outer surfaces may be repaired.

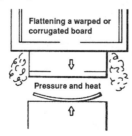

Flattening a warped or corrugated board

Pressure and heat

Handling printed circuit boards

Printed circuit boards are quite expensive (especially multi-layered boards), fragile and easily damaged. Reliability of a product depends to a great extent on printed circuit board quality. Precautions must therefore be taken when handling boards. Main precautions are:

● do not flex

● do not handle

● do not stack

● do not apply pressure to components after soldering

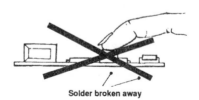

Solder broken away

● grip only by unplated edges.

Important

Cost of multi-layered boards depends on a variety of things such as number of layers, size of board and batch size. However, value of one board may be around 10 to 50 times value of an

**200 by 150 mm (8 by 6 in) one of
a batch of 50
6 layers**

**Approximately ten times the price
of an ordinary printed circuit
board**

**200 by 150 mm (8 by 6 in) one of
a batch of 5
6 layers**

**Approximately twenty times the
price of an ordinary printed
circuit board**

**200 by 150 mm (8 by 6 in) one off
development board
10 layers**

**Up to fifty times the price of an
ordinary printed circuit board**

ordinary printed circuit board. Combined with this is the fact many faults which occur in multi-layered printed circuit boards simply cannot be repaired. Consequently, very great care must be taken of multi-layered printed circuit boards.

Printed circuit board assembly by hand

Tools required
Although other tools may be helpful for specific assembly tasks, tools listed are minimum requirements for a printed circuit board assembler:

1   dressing pliers
2   side cutters
3   tweezers
4   temperature controlled soldering iron
5   snipe-nosed pliers
6   cored solder.

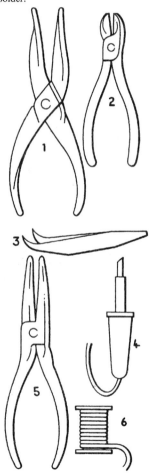

Care of tools

When tools are not in use, pack them neatly in a toolbox. Keep tools clean.

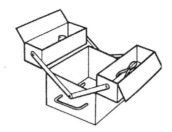

When tools are in use, place them neatly on the bench to be easily reached, and make sure they cannot be knocked to the floor.

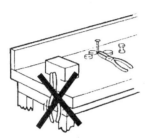

Keep soldering iron in a guard when not soldering.

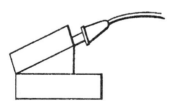

Ensure soldering iron lead is in good condition.

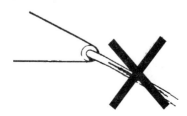

Assembly methods

Methods of printed circuit board assembly depend on components being assembled and type of board.

General assembly requirements

Apart from care in handling printed circuit boards and components, there are many general procedures and practices which should be followed when assembling boards. These include:

● when bending leads of a component ensure specified clearance is achieved. Never bend wires against edges of component end faces. Use a non-metallic bending rod
● alternatively, use a preforming jig to bend component wires

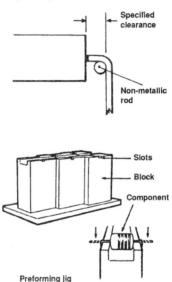

Preforming jig

● when components are mounted on the track side of a board use sleeving and component insulators
● sleeve component leads when they are in danger of short-circuiting against other leads

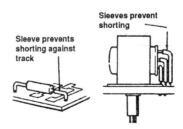

● ensure components are mounted squarely on the board or insulating pads. Leads passing through plated-through holes should not be canted. Do not pull leads through plated-through holes with pliers or similar instruments

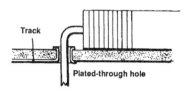

● leads of components which are not heat-sensitive should be cropped and clinched to provide mechanical reinforcement

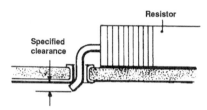

● leads of heat-sensitive components should be cropped to specified lengths then bent to run along the track

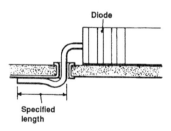

Important

Use a non-metallic tool for clinching and bending. Further:

● use heat shunts when soldering. Always fit a shunt so there is maximum metal-to-metal contact. Do not fit shunt over sleeving. Fit shunt close to component

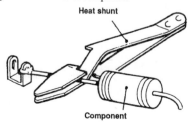

● protect adjacent components by using a heat shield

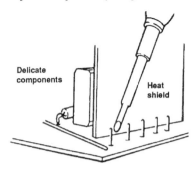

● avoid excessive heat when soldering — this may result in track or pad lifting, or track blistering

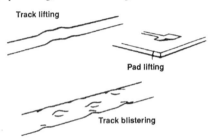

● non-heat-sensitive components may be soldered to completely seal plated-through holes

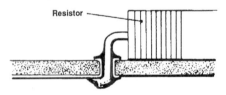

● heat-sensitive components should not be soldered to completely seal plated-through holes

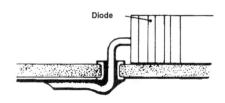

Components must seat squarely on the board. Incorrect seating causes strain on soldered joints.

Component flush with board

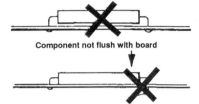

Component not flush with board

There must be no build-up of excess solder on the legend side of the board.

Pad lifted

After assembly boards may be sprayed with protective coating on the track side. This is usually an epoxy resin or varnish.

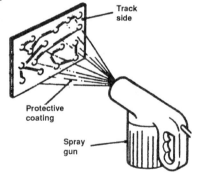

Capacitors
Alternative methods of mounting:

● capacitor body is mounted on the board

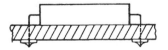

● ensure transistors are correctly identified

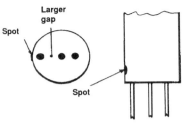

● do not stack assembled printed circuit boards

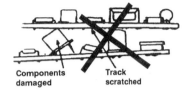

● do not apply pressure to components.

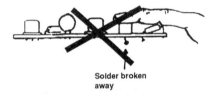

Specific component assembly requirements

Many components assembled to printed circuit boards have unique traits. These must be considered in assembly.

Resistors
There are several ways of mounting resistors:

● special printed circuit resistors with shoulders which rest on the board

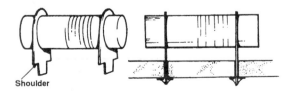

● mounting the resistor body on the board

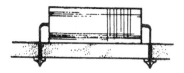

● mounting the resistor vertically, to save space on the board. One end rests on the board, while the exposed lead is insulated if specified

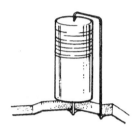

● resistor body is held above the board to aid cooling. Insulator beads may be used to take strain off the solder joints

Insulator beads

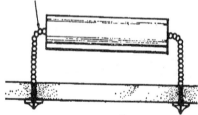

● resistor body is held above the board to aid cooling, while components leads are soldered to pins. In certain applications leads may be formed into smooth bends to absorb vibration — this is known as stress or strain relief.

Pins

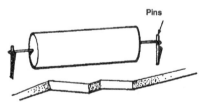

Assembly to plated-through and multi-layered boards

Conventional components such as resistors are assembled plated-through hole and multi-layered printed circuit boards . a similar manner as they are to ordinary printed circuit boards. However, extra precautions and procedures should be followed:

● feed component lead through the board from the legend side
● bend wire end in line with track. Cut wire within periphery of pad

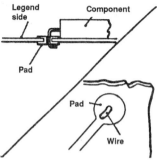

● when hand soldering, use a temperature-controlled soldering iron, at specified temperature. Solder joint as quickly as possible.

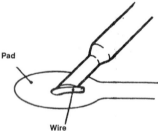

Important

Epoxy glass resin and copper tracks expand at different rates when heated. With multi-layered printed circuit boards, this may cause butt joint failure if overheating occurs on soldering.

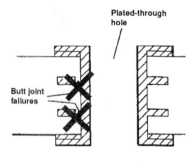

● capacitor body is held above the board to ensure insulation does not break or melt, ruining joints

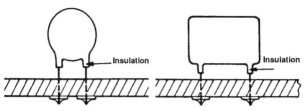

● where special printed circuit capacitors are used, bodies should be mounted on the board.

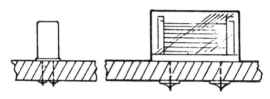

Transformers

Alternative mounting methods:

● transformer body is secured to the board with connection pins soldered to track pads. Insulators may be used between transformer case and the board

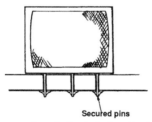

Secured pins

● shroud tags pass through slots in the board and are bent over

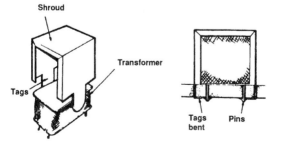

Shroud

Transformer

Tags

Tags bent   Pins

● screws secure body to board.

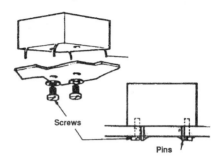

Relays

Special printed circuit relays are mounted on the board. Polarised relays and relays in dual-in-line packages are mounted in connectors fitted on the board.

Discrete semiconductors

Alternative mounting methods:

● transistor body is mounted on the board

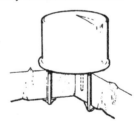

● body is insulated from the board with a plastic spacer. Risk of overheating is reduced with resultant extra lead length

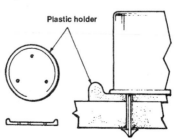

Plastic holder

● body is mounted above the board, with coloured sleeves to insulate and identify leads:

yellow	=	emitter
green	=	base
blue	=	collector
black	=	screen

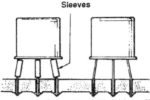

Sleeves

● body is mounted horizontally and held in a clip.

Use a copper heatsink to aid dissipation of heat from the transistor and prevent overheating where specified.

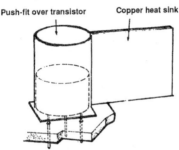

Push-fit over transistor   Copper heat sink

Permanent heatsinks are often fitted to dissipate heat from power transistors. They usually comprise finned structures machined from aluminium castings or extrusions. To mount a transistor on a heatsink:

● ensure heatsink surface is flat and clean. Remove burrs from holes
● if appropriate use a mica insulator between transistor body and heatsink

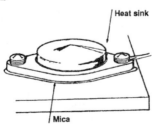

Heat sink

Mica

● coat both surfaces of mica insulator with heatsink compound, to aid heat transfer between transistor and heatsink.

Thin-film integrated circuits

Mounting precautions and procedures include:

● pins on thin-film modules are set in ceramic material which is easily cracked. Pins may break, too

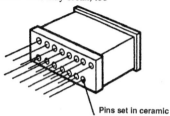

Pins set in ceramic

● pins are usually in a formation preventing incorrect insertion

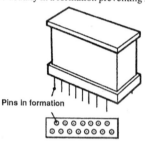

Pins in formation

● legends should correspond to pin formation

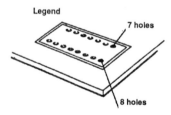

Legend

7 holes

8 holes

● straighten pins, if necessary, before fitting. Take care not to break pins

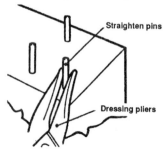

Straighten pins

Dressing pliers

ly requirements
ins close to board. The
ng.

● locate pins in holes then push them through — gently

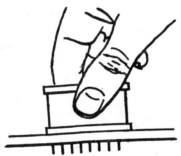

● to secure components, bend and tack solder three pins. Take care not to damage tracks and pads when bending pins

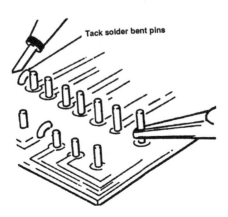

Tack solder bent pins

● straighten tack soldered pins. Cut all p
component is now ready for wave solder

Important
Components must be seated flat to the board.

Components must be fitted tightly. No sideways lateral movements are allowable.

TO5 packaged integrated circuits
Procedure for assembling TO5 packaged integrated circuits to
printed circuit boards includes:

● pass device pins through a pad so they protrude through pad
underside

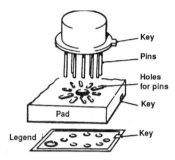

● locate pin ends in correct holes on the printed circuit board and
push the device down firmly

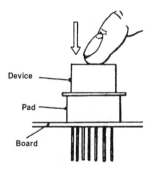

● hold the device in position with a finger while bending two
pins, then tack solder them to hold the device

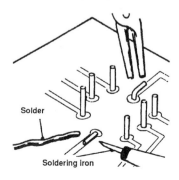

● straighten tack soldered pins and cut all pins close to the board.

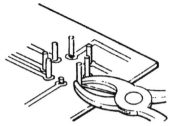

The device is now ready to be wave soldered.

Important
Devices must be seated flat to the board.

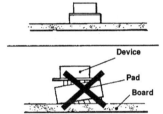

Device key must correspond to pad key and legend key.

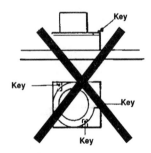

Device must be fitted tightly, with no movement.

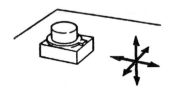

Dual-in-line packaged integrated circuits

Procedure for assembling dual-in-line packaged (DIP) integrated circuits to printed circuit boards includes:

● correct arrangement of pins before fitting device to printed circuit board. This can be done in a special jig. Line up device pins over location holes in the jig, then

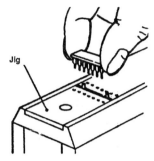

Jig

● push down firmly. A spring release ejects the device

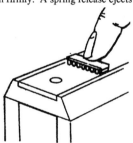

● line the device pins up with holes in the printed circuit board ensuring device polarity mark and legend correspond

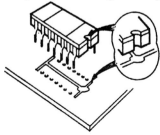

● fit device carefully, ensuring pins insert fully

● ensure device is flat to board

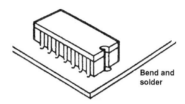

Bend and
solder

● secure device by turning two leads outwards. Tack solder these two leads, and cut all leads close to board. Device is now ready for wave soldering.

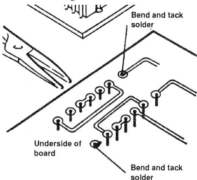

Bend and tack
solder

Underside of
board

Bend and tack
solder

Important
Devices must be seated flat to board.

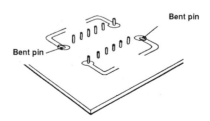

Bent pin

Bent pin

Only turn the specified number of pins outwards to secure the device.

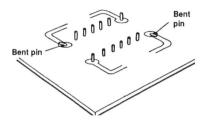

Device polarity mark and legend must correspond.

Devices must be fitted tightly, with no movement.

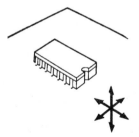

Flatpack integrated circuits

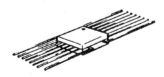

Flatpack devices can be assembled in the conventional method, that is their pins go through the printed circuit board and are wave soldered. In this case:

● pins are preformed by jig

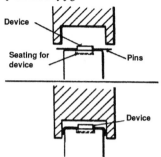

● device is fitted, tack soldered then wave soldered to the board.

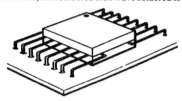

An alternative assembly method is:

● device is glued in position on printed circuit board. Device leads must overlay the track exactly

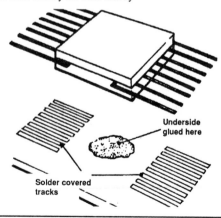

Important
With this method, flatpack devices are assembled after all other components have been wave soldered. Ensure glue does not spread onto track.

● solder pins by a reflow method such as hot air (shown), infra-
red rays, hot oil, electrical pulses, laser.

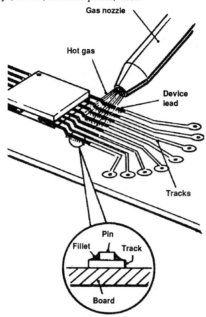

Flatpack integrated circuits can be hand soldered to a printed
circuit board:

● place device in position and hold
● secure device by running a soldering iron along two or three
device pins
● solder remaining pins.

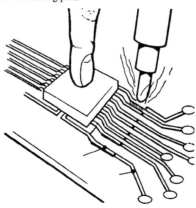

Important

Soldered track and device pins must correspond.

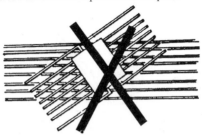

Pins must not be lifted. They must be fully attached to track.

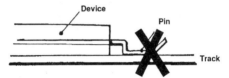

Soldered joints must be formed with a good fillet.

Hand soldering

When soldering, use of correct tools, materials and techniques is essential and care must be taken to avoid burns, fire, or electrical shock, and to avoid damage to components and equipment.

Electrical maintenance of soldering irons
All connections should be examined regularly and kept clean and tight. Mains lead must be replaced immediately if insulation is burned or chafed.

Mechanical maintenance of soldering irons
When in use, an iron's bit must be kept clean, using a wire brush, abrasive pad, or damp sponge. Working surfaces must be kept tinned. Regularly, when the iron is cold, rotate the bit in its housing to free corrosion which may otherwise lock the bit and prevent its removal.

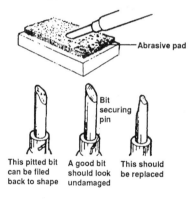

— Abrasive pad

Bit securing pin

This pitted bit can be filed back to shape

A good bit should look undamaged

This should be replaced

Care in use
Irons must always be:

● replaced on stands — stands must be kept well away from bench edge
● switched off after use
● allowed to cool before putting away.

Burns
Take care not to burn hands when soldering in a restricted space.

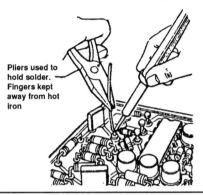

Pliers used to
hold solder.
Fingers kept
away from hot
iron

Safety
There are many important points regarding safety when using
soldering irons:

● electrically isolate the equipment before soldering
● check electrical connections to iron and plug regularly
● take care an iron does not burn through its own lead
● never allow an iron to come into contact with flammable
material
● replace the iron carefully on its stand
● remember joints remain hot for some while after soldering.

Types of solder
Soft solder is an alloy of tin and lead. It is used to join metal
together as it melts and flows onto two surfaces to provide a film
uniting them. Note soldering does not melt the two surfaces,
merely combines them with a melted metal, which solidifies on
cooling.

Flux
cores

   To aid flow of molten solder onto the two surfaces to be
joined, flux is used. Solder used in hand electronic work is
usually in a cored form, with a core or cores of flux. Flux is
sometimes used in paste form, painted onto a joint prior to
soldering, but this is not so convenient. There are two main
types of flux:

● resin flux — which has two main functions; (1) to flow quickly over the joint, speeding up transfer of heat to the work; and (2) to protect the metal surfaces from contamination, either by dirt or metal oxidation

● active flux—which has the functions of resin flux, plus it cleans joint surfaces by removing grease and oxidised film by corrosive action.

Cores of cored solder are usually filled with resin flux, although solder with cores of active flux are available.

Types of soldering iron
Correct size of iron is critical. Too large an iron is clumsy to use and may cause component damage with excessive heat. Too small an iron, on the other hand, will not produce enough heat to bring large components to a high enough temperature.

Heavy iron
A heavy iron is necessary for heavy work, such as steel chassis or connections to heavy cables, in order work is heated sufficiently and joints can be made rapidly.

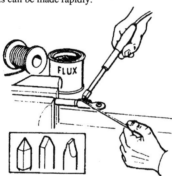

Medium iron
Used for most work involving installation and removal of components.

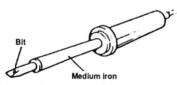

Sub-miniature iron
Intended for precision work with dedicated components and printed circuit boards.

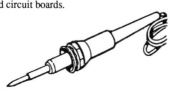

Soldering gun
Soldering guns are frequently specified for intermittent operation in general-purpose applications. They heat up to working temperature rapidly.

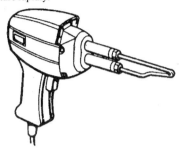

Regulated irons
Regulator units are used to control soldering iron bit temperature.

The electric soldering iron
An iron's heating element is heated by electric current passing through it. The bit is heated by the heating element. Face of the bit is the part of the iron which makes contact with surfaces to be soldered.

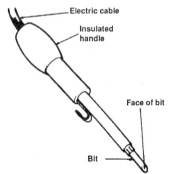

Electric cable

Insulated
handle

Face of bit

Bit

A soldering iron is designed to operate at a specified voltage.
This can be either:

● mains supply
● a low voltage supply.

Soldering irons are available in a range of power ratings. Power
is specified in watts.

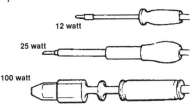

Irons should be selected with an adequate power rating for
size of work.

Most bits are made of copper as it is a good conductor of heat.
Bit faces, on the other hand, may be either:

● unplated
● iron plated.

Iron plated bit faces do not wear as rapidly as unplated bit faces.
Most irons are constructed so bits can be changed. Different
bit shapes are usually available.

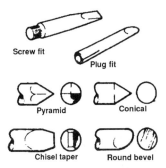

Select the bit to give the best compromise between:

● best approach to work
● shortest bit and bit taper
● best contact with surfaces to be joined.

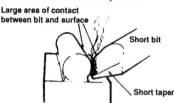

Large area of contact between bit and surface

Short bit

Short taper

Unplated bits quickly become pitted, mis-shaped and covered in oxide. If an iron is in constant use, this occurs within just a few hours. Such unplated bits must therefore be dressed regularly in a simple process:

● switch off iron, unplug and allow to cool
● remove bit from iron

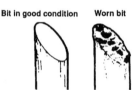

Bit in good condition   **Worn bit**

● mount bit in a vice — not in an electronic assembly area, as copper dust from the bit may settle in equipment and cause short-circuits
● file to shape.

Iron plated bits must not be filed.

Both types of bit must be cleaned regularly in use. This is a simple procedure of wiping the hot bit over a damp sponge pad (plated) or wire brush (unplated). Note unplated bits must not be wiped over a wire brush.

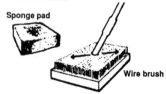

Sponge pad

Wire brush

Joints

To make a good joint, solder must flow evenly over surfaces to be soldered. How evenly solder flows is usually described by the term wetting. Good wetting occurs if:

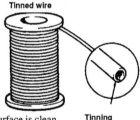

- surfaces are clean
- sufficient flux of correct type is used
- surfaces are hot enough
- surfaces have been tinned.

Tinning untinned surfaces

Most surfaces to be soldered will have been pre-tinned. However, to tin untinned surfaces:

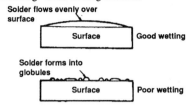

- ensure surface is clean
- heat surface with iron
- apply solder
- remove solder, then iron.

Tinning a wire

Insulation must first be removed from a wire, and it must be cleaned. Multi-strand wire should be twisted. Then:

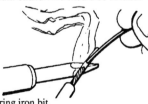

- tin soldering iron bit
- rest wire on bit face
- apply solder to wire without solder touching the bit
- apply solder to all areas of wire which need to be tinned
- withdraw solder and lift away the tinned wire
- confirm the solder film is clean and shiny.

Note it is essential the tinning process is accomplished in the minimum of time.

Mechanical reinforcement

This is the process of connecting wire prior to soldering. Application of solder ensures electrical conductivity, but does not guarantee strength. Strength of a joint depends almost entirely on mechanical connection. While component joints on a printed circuit board may be strong enough to adequately hold minute components, this is not the case with larger components or wire connections. Such connections must always be tight otherwise the joint may break down in service.

Metal terminals, mounted on insulating material, are used to make soldered joints at wire junctions and to secure components. Main terminal types include:

- plain terminal posts (1)
- plain tag terminals (2)
- turret terminals (3)
- flat terminals (4)
- bifurcated terminals (5)
- hook terminals (6).

Terminal boards or tag boards may be used to wire junctions or to support components.

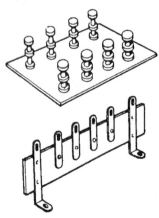

Wire wrapping
Method of wire wrapping depends on terminal type, although all methods follow a similar pattern. Degree through which wire is turned, and clearances required, are usually specified in company procedures or practices.

Plain terminal post
When wrapping wire around plain terminal post:

● hold wire against post

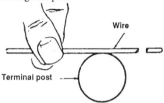

● grip wire gently with pliers and bend

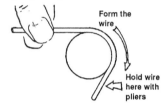

● rotate pliers to form an outward bend

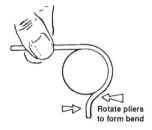

● cut off excess end of wire with side cutters

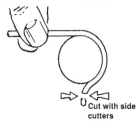

● pinch wire lightly with pliers and ensure wire is secure.

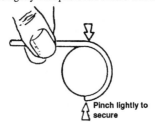

Pinch lightly to secure

Tag terminals

Two methods may be used here, according to mechanical reinforcement required:

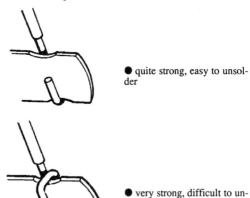

● quite strong, easy to unsolder

● very strong, difficult to unsolder.

Flat terminals

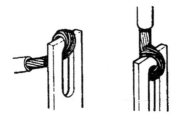

These terminals are used on terminal strips, switches, cable connectors and so on. Wrapping technique depends on wire direction.

Turret terminals

These are occasionally used to support components. Component leads must be bent following the plain terminal post wire
wrapping procedure. Sharp tools should never be used to start
a bend. Never bend leads against component ends and all bends
must be smoothly formed.

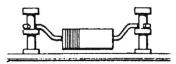

Bottom portion of a turret terminal supporting a component
may be used to secure interconnecting wiring.

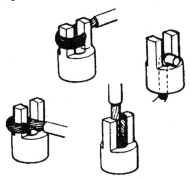

Bifurcated terminals

Due to their shape, these terminals provide various methods of
wrapping wire.

Hook terminals

There is only one method of wire wrapping a hook terminal.

Insulation gap

It is important when wrapping wire to leave a sufficient gap
between joint and wire insulation, otherwise heat of soldering
may damage insulation.

**Insulation can melt if too
small a gap is left**

Making a soldered joint

Tinned copper connecting wire of a resistor is to be soldered to
a pre-tinned solder tag.

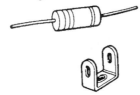

Tools and materials needed are:

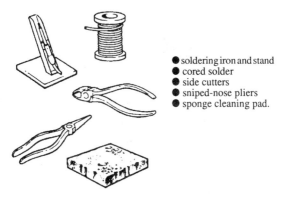

- soldering iron and stand
- cored solder
- side cutters
- sniped-nose pliers
- sponge cleaning pad.

To prepare tag, insert component lead through hole. Using
pliers, bend the lead horizontally round until it completely
encircles tag. Cut off excess wire.

When preparing soldering iron ensure:

● iron has no physical damage and is of correct voltage and power rating
● bit is of suitable shape and in good condition
● wire and tag are clean
● connect the iron to its power supply, switch on and allow to warm up.

Hold iron comfortably. Rest an arm on the bench if desired, to reduce fatigue and improve hand stability.

Hold iron in a pencil grip

To tin the bit:

● rub bit on cleaning pad to clean
● apply solder to bit. If bit is not completely and evenly covered with solder, clean and apply solder again
● wipe bit gently on cleaning pad to remove excess solder.

Safety
Never flick excess solder off the bit. Hot solder burns. Hot solder may fall into work and cause a short circuit.

When soldering, check iron is at correct temperature. Irons take 2 to 10 minutes to heat up fully, depending on power and size. At correct temperature solder should melt immediately to a shiny surface, flux does not char but should give off a little white smoke.

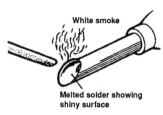

Wipe iron clean, then use the flat of the bit to press solder onto the joint. As solder melts, feed in more solder until joint is fully covered. Remove solder and iron. Do not move components until solder is seen to solidify. Allow joint to cool naturally, do not blow cool.

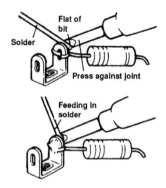

Remove solder before removing iron, to avoid solder spikes. Remove iron as soon as solder flows over all joint surfaces.

Multi-connections may involve several conductors at one point, so should all be positioned before final soldering. Multi-wrap joints are used when connecting fine wire ends to pins. Open slot tags are used if components are likely to be changed.

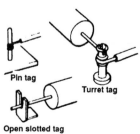

Soldering ideals and faults

A good joint has a very low electrical resistance. A bad joint may have a low initial resistance but this may not be maintained.

Correct amount of solder

Solder must flow freely around a tag and its connection. Contour of connection wire should be plainly visible through solder.

If joint is dirty, solder does not flow over dirty parts and a dry joint results.

Excessive heating may damage:

● wire and insulation
● component
● adjoining components.

Do not overheat wire. If this happens solder flows up the wire and makes multi-strand wire brittle. Movement of solder up a wire is called wicking.

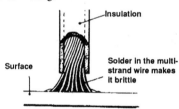

Insulation

Surface

Solder in the multi-strand wire makes it brittle

Some common soldering faults are illustrated:

● excess solder (blobby)
● joint too low, also splashes of solder

● overwrapped and spiky
● sleeving too near to joint

● sleeving too far back
● solder flows down self-fluxed wire

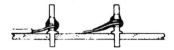

- loosely wrapped joint
- tacked on joint.

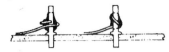

Printed circuit assembly and soldering

When hand soldering components to a printed circuit board a soldering fixture is normally used. This comprises a rectangular wooden box filled with a sponge rubber pad. A metal lid, hinged to the box, has a rectangular hole in it.

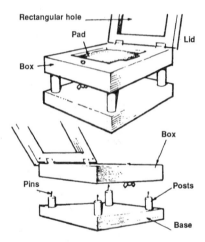

Four posts protruding from a base have pins on top, and the posts can support either wooden box or lid.

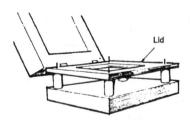

Component assembly takes place with printed circuit board supported on metal lid.

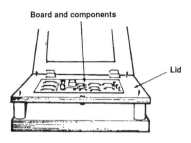

Soldering takes place with printed circuit board supported on wooden box.

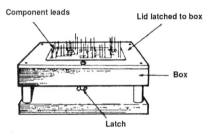

Order of component mounting and soldering

Although there is no strict procedure for component mounting and soldering, task is basically a repetition of a single main process:

● support metal lid of fixture on the four posts, and rest printed circuit board on rectangular hole edges, track side down

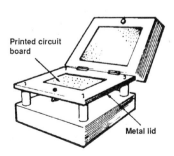

● insert all resistors into correct positions in printed circuit board. Ensure they fit flush to the board. Ensure all colour codings on all parallel resistors read in one direction

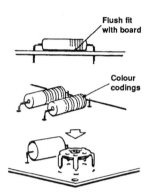

● insert all similar height components (eg, preset resistors)
● close box onto metal lid and latch in position. Components are now clamped and held steady by sponge rubber pad

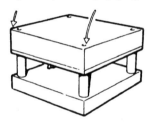

● reverse fixture and support box on posts

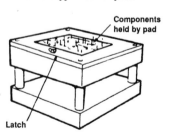

● solder each component lead to its pad

● remove excess lead ends

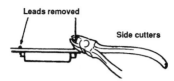

● repeat procedure with next highest components until printed circuit board is fully populated and soldered.

When component mounting and soldering is undertaken by hand, components are assembled in an order specified in a sequence chart. This lists component order and any special instructions.

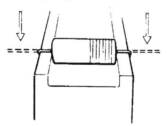

A preforming jig is often used to preform component leads to a specified shape. Operation is simply a matter of holding component body on jig, then stroking leads down with fingers, until leads are at right angles to component body.

Component position is often shown on a component layout drawing, or may be indicated by printed circuit board legend.

Just before inserting a component, clean its pads with a soft rubber eraser — do not rub too hard — and clean its leads. Then fit the component.

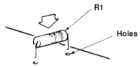

Component soldering

Generally, though by no means always, components are mounted on the opposite side of a printed circuit board to its track. This cannot, however, be the case for double-sided printed circuit board. Component leads pass through holes in the board and are soldered to track pads.

Good component joints on a printed circuit board can be made in a number of ways, including:

● component lead end is at right angles to printed circuit board

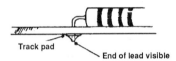

● lead end is at an angle of less than 30° from vertical

● lead end is bent parallel to printed circuit board, in direction of track.

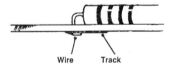

In all these good joints, however, end of the component lead is visible through solder and solder surface is concave.

Printed circuit board soldering faults

Faults can occur when soldering components to printed circuit board in a number of ways, including:

● spiking

● excess solder bridging across track or pads causing short circuit

● excess solder

● excess flow down contact

● excess solder, outline of lead not visible through solder

● excess solder, build-up on track and areas without holes

● ring of dirt round lead and dull appearance of solder, possibly blow hole

● insufficient solder

● poor solder flow, appears dull and blobby

● dry joint, possibly dirty lead.

Soldering delicate components

Any component can be overheated while soldering, by conduction of heat along wires: small components are particularly

susceptible. Although transistors and semiconductors may show no external signs of damage, overheating may cause an internal short or open circuit, or just impaired performance. If a glass seal is cracked, a semiconductor component deteriorates slowly.

Resistors are less easily damaged. Painted rings used for colour coding char and smoke when overheated. Resistance value may be altered.

Small capacitors of plastic encapsulated types melt when overheated. Breakdown may occur as a result.

Insulating materials blister, char, melt or shrink.

Tags and pins are loosened and track becomes detached from printed circuit boards.

Precious metal coatings may be damaged or removed.

Heat shunts

When soldering small components, particularly those known to be heat-sensitive, a heat shunt should be used. A heat shunt is clipped onto the component lead during soldering. Being a good heat conductor it diverts heat from the component and prevents heat damage to the component. The shunt is removed only when a joint is cool.

Best heat shunt is one which can be clipped to a component lead between joint and component, as it leaves both hands free to solder.

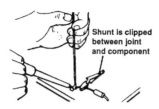

Shunt is clipped
between joint
and component

Pliers can be used effectively, but need to be held while soldering. Care must be taken not to disturb the joint while it cools.

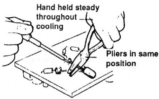

Hand held steady
throughout
cooling

Pliers in same
position

It is important to obtain maximum metal-to-metal contact. Shunts must not be fitted over sleeving. If possible, shunt must be fitted at an angle close to the component.

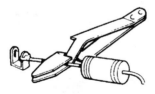

Shunts should be fitted to all component leads connected to a joint — one shunt for each lead.

Reflow soldering

Soldering individual components by machine is often undertaken in a reflow process. Such a process (there are many types)
involves tinning surfaces to be soldered before assembly. Surfaces are then held together and heated until solder tinning flows
again (hence the name reflow), at which point heat is withdrawn
and the joint solidifies.

Often reflow soldering is undertaken with a heating element,
which also serves to hold surfaces together. Current is passed
through the element for a predetermined time, after which
pressure is maintained until joint and element cool. Then the
element is lifted automatically.

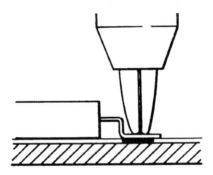

When setting a reflow soldering machine ensure pressure is
set as specified: too little pressure allows air into the joint; too
much forces solder from between the surfaces.

Solder pre-forms

A version of reflow soldering can be undertaken by hand,
usually where access is difficult, using solder pre-forms to form
joints. Solder pre-forms of various shapes, diameters and
thicknesses are available.

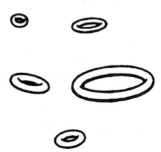

Process may be speeded up or improved by preheating both parts of the joint. Two steps only are required:

● place the pre-form in position between the two surfaces

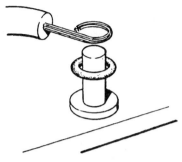

● apply pressure on the solder ring and heat using a hot air blower, or a soldering iron, until solder pre-form melts and flows to make a joint. Remove pressure when the joint is cool.

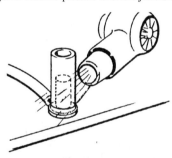

Components, piece parts and sub-assemblies

Printed circuit boards
For rapid identification, colour splashes are often used in addition to part numbers. Some boards have handles and these are often coded.

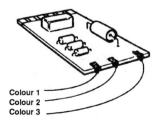

Colour 1
Colour 2
Colour 3

Composite printed circuit assemblies

Composite assemblies of printed circuit boards are made up by mechanically fixing:

● two similar boards to form one assembly
● by mounting smaller printed circuit boards on a larger board
● by mounting a large number of small sub-boards or daughter boards onto a large main or mother board. This arrangement also has other names, such as a tank or nest of printed circuit boards.

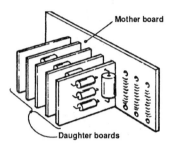

Sub-boards can be mounted on a main board to lie parallel with it. Electrical connections are made with plug and socket connectors, arranged to allow sub-boards to be stacked. Sub-boards are removable.

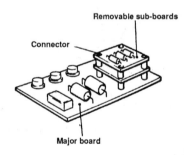

Connectors can be used to mount a sub-board vertically on a main board. In this case sub-boards can be stacked side-by-side.

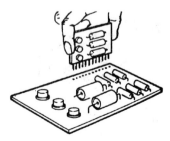

Sub-boards can be permanently attached to a main board with nuts and bolts, with spacers maintaining required distance between boards. This type of mounting allows wired connections between boards. Care must be taken to ensure nuts and bolts are tightened evenly and not over-tightened — this may crack boards.

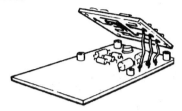

Two boards of same or similar size can be bolted together with spacers to form a composite assembly. Resultant two edge connectors often mate with a double-socket connector.

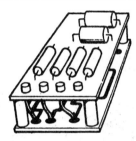

A number of similar boards can be mounted together in a folded stack. This system allows boards to be close-packed while allowing easy access for servicing.

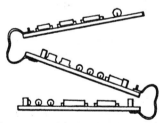

Flexible connectors join boards. These connectors allow boards to be physically separated while electrically connected. While stacked, boards are bolted together.

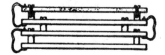

Edge connectors

Board end connectors are an integral part of many printed circuit boards, with track continued to board edge. On some boards track is wrapped right round the board edge.

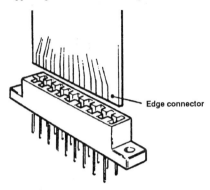

Edge connector

Wire pins

Body of a wire pin-type connector is fixed firmly to the printed circuit board. Wire pins are soldered directly to track, but the body permits no flexing or strain on joints.

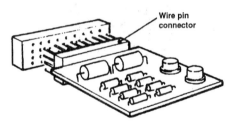

Wire pin connector

Tag connectors

These provide a separate removable tag for each circuit. Their extremely small size allows many circuit connections in a small space.

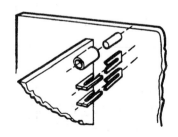

Encapsulation of printed circuit boards

Printed circuit boards are sometimes encapsulated with a protective layer of insulator, known as a conformal coating. This is a thin transparent coat of acrylic or epoxy-type material which helps protect printed circuit boards against humidity, dirt, vaporous contaminants and foreign bodies such as metal filings. Conformal coating does not however, as is commonly thought, help a printed circuit board withstand vibration. Encapsulation by hand is a straightforward process:

● mask all connectors eg, edge connectors, spills

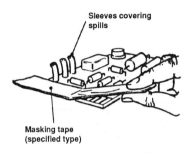

Sleeves covering spills

Masking tape (specified type)

● clean the board with an approved cleaning agent. Brush correct primer over the board. Ensure all surfaces are covered with primer. Leave to dry

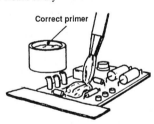

Correct primer

● mix encapsulation compound, then brush onto printed circuit board. Ensure all surfaces are covered. Leave to dry. Note the compound, after mixing, remains workable for a limited time only, depending on chemicals used. Work quickly so this limit is not reached while any one board is being coated

● after board coating is set, place board in encapsulation tank and bake for specified time to cure

● when cured remove board from tank and check coating is smooth, well covered and unbroken. Remove masking from connectors.

Repairs to printed circuit boards

Before commencing any repair to a printed circuit board, check board specifications allow the type of repair required. Some specifications forbid certain repairs, and so these should not be attempted.

Broken track
Where a track is broken but remains firm on the base of the board, a strip of material similar to the track may be soldered across the break:

● clean track and bridge before repair
● when cool, cover the joint with an acceptable varnish.

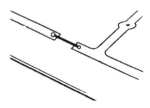

Lifted track

Where a track lifts in one place but otherwise adheres well, it is sometimes possible to repair:

● clean board and place it, component-side down, on a piece of sponge rubber or similar protective material so components are not damaged while work is carried out. Carefully cut across track on either side of lifted portion

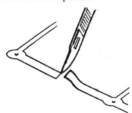

● remove cut piece leaving a gap in the track

● drill four holes, two on each side of the break. Check hole positions prior to drilling, so the drill does not damage components

● form wire bridges and insert in drilled holes

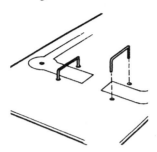

● insert new portion of track under bridges. Bend bridge legs on board component-side to hold the new portion of track firmly. Solder bridges, new track and old track.

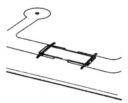

Damaged or lifted pads

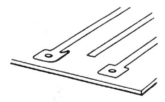

Track pads sometimes lift and may break away. It is often possible to repair such faults:

● remove any component connections. Use a soft brush and a safe acceptable cleaning fluid to clean underneath lifted pad, or area of missing pad. Fix lifted pad with suitable adhesive

● for a missing pad, form new pad from similar material to original pad. New pad should have a longer tail than the missing piece. Using acceptable adhesive fix the new pad in position, working the longer tail under the remaining track

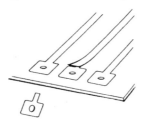

● work pad gently but firmly to ensure no air is left under track. Leave adhesive to set before resoldering component and varnishing joint.

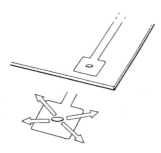

Components on an encapsulated board

Replacement of faulty components on an encapsulated board is only possible if a pliable encapsulating material is used. Procedure is:

● using a sharp blade make cuts in the encapsulate around the faulty component. Cuts must be deep enough to break the seal but not deep enough to damage the board

● gently peel the encapsulate from component, making sure other components are not stressed. Desolder and remove faulty component

● fit and solder replacement component. Inspect and test board for correct operation

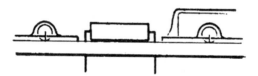

● re-encapsulate the area stripped, following same procedure as for full encapsulation. Re-inspect and test board.

Fitting components to front panels

Front panels of chassis or equipment normally have a high
quality finish. Most work involving fitting of components to
front panels deals with how to maintain such a finish. Prior to
drilling or punching front panels to allow components to be
mounted, check, re-check and check again holes will be in the
correct place.

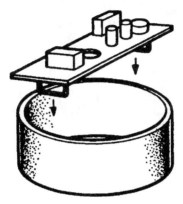

Treat front panels carefully — do not scratch or mark them.
Use a roll of tape or similar means to protect polished surfaces
while work is carried out.

Potentiometers may be mounted on two pillars. Use standard
flat washers to protect surface of panel from damage by screw
or bolt heads.

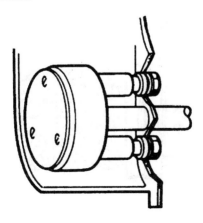

Centre-mounted potentiometers and switches need only one large hole. Set potentiometer or switch body in the correct position and hold it steady while tightening the nut with a box spanner.

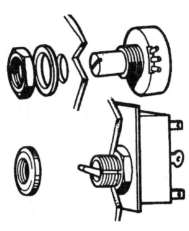

Potentiometers
End stops of potentiometers are often fragile. Do not use excessive force to turn the spindle. Keyways may be used to ensure correct orientation.

Switches
Screw on the knurled nut finger-tight. Secure assembly by tightening rear nut with a spanner.

Before mounting wafer switches verify from drawings correct orientation of each wafer — they can easily be assembled with a 180° error. Examine each tag, straighten bent ones and replace broken ones. Drill out damaged tags and rivet new tags in place.

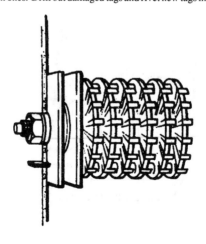

Connect any inter-unit wiring and components, allowing space for subsequent leads on each tag if required, before installing switch. Wire carefully and neatly to prevent damage to wafers and tags and to aid checking. Test the operating mechanism and ensure it is clean, brisk and positive. Compare wiring circuit and switch diagrams to ensure they agree with each other and with panel markings.

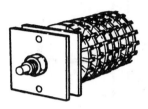

Fuses
Cartridge or cylindrical fuses are often provided on a front panel for easy access. Fit the body of the holder through the front panel and secure with locking ring. Check fuse provided is the correct rating. Check the cartridge is held firmly in the cap and spring in the body is effective.

Variable capacitors
Variable capacitors may be mounted on a chassis with only the spindle protruding through. Slow-motion drives are often used, which may be self-contained in a large-sized knob. Set knob position in relation to capacitor vanes and front panel markings, before locking knob firmly to shaft.

In certain cases the control knob is on a dummy spindle which carries a pulley. The pulley is mounted behind the front panel and not on the panel itself.

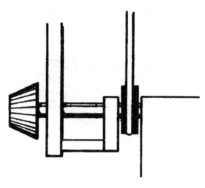

Trial-fit major components such as multipole multi-wafer switches and check spindle projects far enough for a knob to be fitted.

Knobs

Knobs on front panels are fixed to spindles of rotary switches, capacitors and potentiometers in a number of ways. Knobs must be easily removable for servicing purposes. Most types use a knob fitted with a grub screw which screws onto the side of the component spindle.

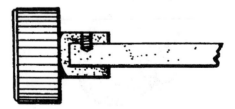

Another type has a grub screw in the top of the knob under a cap (screw A). The end cap is removed before the knob is fitted.

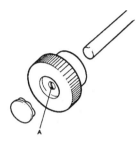

Indicators

Ensure colours of glass in indicator lamps are correct. Bodies are secured either by a nut or by the cap. Many indicator lamps use only one connection, chassis earth used for the second connection. For such indicators, it is important to ensure a good connection to the chassis by removing any paint from the inside surface around the hole prior to assembly, and using a spring tab washer behind the panel.

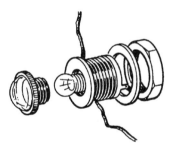

Meters

Meters are delicate instruments and can be damaged by mishandling. Protect movements of low current reading meters with a shorting link across terminals when not connected in circuit.

When handling and fitting meters ensure glass is not scratched.

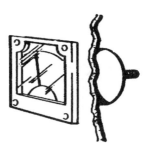

Meter cases are often fragile. Do not overtighten holding screws.

Fixing labels to panels and racks
Labels may be fixed by:

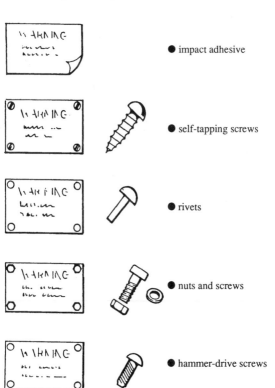

- impact adhesive

- self-tapping screws

- rivets

- nuts and screws

- hammer-drive screws

Appearance of a front panel is important. Always check components are fixed correctly and neatly. Check all visible washers are of the same size. Check no screws are burred and the panel itself is not scored during assembly. Protect finished surfaces when doing work. Tape handles, meter cases and all projections and edges.

Fitting components to chassis

Certain types of printed circuit board may be bolted directly to brackets or ledges on a chassis. Connections are made by direct soldering to track. Do not overtighten fixing devices; the board material may crack.

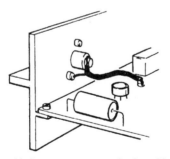

When soldering components terminating with solder tags, ensure leads to be connected cannot distort or damage components. Tags are not meant to take large cable weight, so ensure cables are not free to move.

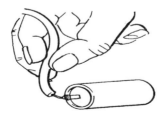

If necessary provide mechanical support with a clip.

Components can be supported with saddles. In such a case ensure values and other written details on component bodies are not obscured.

Observe component polarity where applicable.

Use anti-vibration mountings to protect sensitive components, such as variable capacitors. Fit a washer of elastic material between component body and chassis. Tighten fixing device, usually a nut and bolt, until fixing is firm and washer is slightly distorted. Use an approved method to lock nut and bolt assembly.

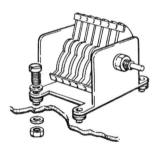

Use a washer of harder material if mounting is very soft, or washer is very distorted. Use a washer of softer material if fixing tightens without distorting washer at all.

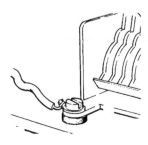

Note anti-vibration mountings insulate component body from chassis. Ensure, where required, a good earth connection is made. Often earth connections can be made with flexible braid.

Anti-vibration mounts are used to support heavy items such as fans and motors. In such cases, of course, they prevent vibration from component to chassis (rather than the other way round).

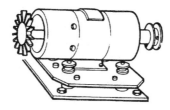

Where larger components are mounted, ensure a good metal-to-metal contact. Clean mating surfaces to remove paint, varnish or encapsulate. Seal the completed joint with a varnish or specified means.

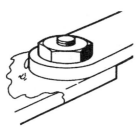

Intimate contact of dissimilar metals
When two dissimilar metals are in contact a potential difference develops between them. A rule-of-thumb is: if this is greater than 0.5V (0.25V in the tropics) corrosion may occur.

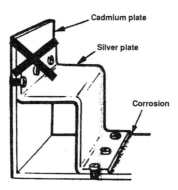

Cadmium plate

Silver plate

Corrosion

Table of contact potential differences (volts)

	Gold	Rhodium	Silver	Nickel	Copper
Zinc	1.40	1.40	1.11	0.96	0.92
Cadmium	1.08	1.08	0.79	0.64	0.60
Aluminium	1.05	1.05	0.76	0.61	0.57
Steel	1.00	1.00	0.71	0.56	0.52
Chromium	0.80	0.80	0.51	0.36	0.32
Tin	0.77	0.77	0.48	0.33	0.29
Stainless steel	0.75	0.75	0.46	0.31	0.27
Brass	0.60	0.60	0.31	0.16	0.12
Copper	0.48	0.48	0.19	0.04	0.00
Nickel	0.44	0.44	0.15	0.00	
Silver	0.29	0.29	0.00		
Rhodium	0.00	0.00			
Gold	0.00				

For example, a silver-plated bracket must not be screwed directly on a cadmium chassis as a potential difference of 0.79V is created.

However, if tinned washers are introduced between bracket and chassis, potential differences are 0.31V between tin and cadmium, and 0.48V between tin and silver. Both are within the limit.

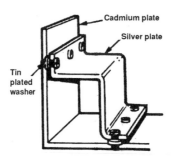

Cadmium plate

Silver plate

Tin plated washer

Encapsulated and sealed relays usually fit into standard chassis punchings. Insert relay into aperture, ensuring it is correctly orientated. Insert correct screws with spring washers into tapped holes, and tighten.

Brass	Stainless steel	Tin	Chromium	Steel	Aluminium	Cadmium	Zinc
0.80	0.65	0.63	0.60	0.40	0.35	0.32	0.00
0.48	0.33	0.31	0.28	0.08	0.03	0.00	
0.45	0.30	0.28	0.25	0.05	0.00		
0.40	0.25	0.23	0.20	0.00			
0.20	0.05	0.03	0.00				
0.17	0.02	0.00					
0.15	0.00						
0.00							

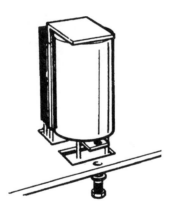

Relays with open contacts are particularly susceptible to damage through poor storage, transit and handling. Before fitting a relay, examine carefully all leads and contacts.

Straighten slightly bent leaves. Do not attempt to rectify damaged leaves.

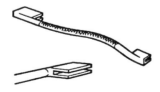

Where units are rejected in test due to relay malfunction, examine contact pads. If dirty or pitted, refurbish using a special contact file.

Servo components used in electronic work comprise small rotating machines with related gear trains and electrical circuit. Machines (synchros, magslips, M motors and so on) are normally end-mounted with spindles protruding through the mounting face.

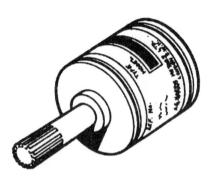

To fit into an assembled gear train, gently insert splined shaft until it meshes with the gear. Turn servo body to bring fixing holes in line, and insert.

Fix machine to its fixing plate, keeping spindle in alignment by tightening each screw in order, a few turns at a time, until fully home.

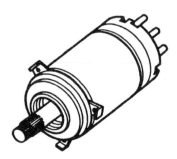

These components are adjusted by rotating their body in relation to shaft and mounting. Ensure each is mounted so body locking screws are accessible.

When wiring up, ensure sufficient slack is allowed in wiring to permit 180° rotation.

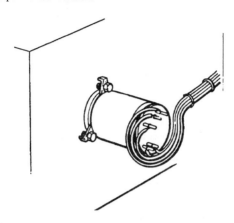

Preset components

IF and RF transformers and inductors are often tuned by screwing a ferrite core in or out of the hollow centres of coils. Any movement of the core after setting detunes the circuit, so must be prevented.

Note cores in such components are normally only adjusted by technical department staff.

Methods to limit core movement include:

● a bush or double-wedge of nylon or other suitable material is inserted in the threads. This absorbs external vibration and shock, damping core movement

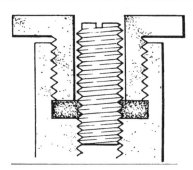

● a split washer is sprung across the top of the hollow centre, and the core protrudes through the split. It is prevented from turning by frictional pressure of spring against thread. If adjustment is necessary, downward pressure releases spring tension

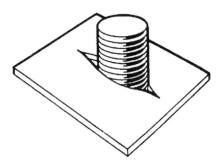

● a rubber band stretched across component top acts in much the same way as a split spring

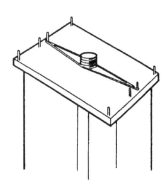

● melted wax is poured into the centre hole. After it solidifies the core is held firmly. As soft wax is used it can be easily removed if adjustment is required.

Small preset capacitors and potentiometers are often provided with a locking nut. Tightening this nut closes the split jaws of a core which grips the spindle. Set the capacitor with a screwdriver and hold the spindle steady while tightening the nut.

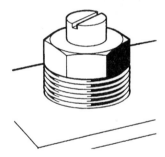

Setting of such a component must always be checked after locknut tightening.

    Coat all external preset controls after setting with an approved varnish or paint. Use coloured varnish or paint where possible, as this aids visual confirmation of locking.

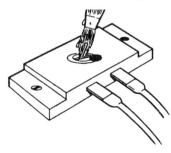

Prototype assembly

A development technician or prototype assembler must often produce an assembly, working only from a circuit sketch. Jobs such as this are normally carried out in consultation with a development engineer. There are typically two main jobs to be undertaken when building prototypes:

● circuit board assembly
● chassis assembly.

Typically, prototype assemblies are constructed in a rack-mount chassis.

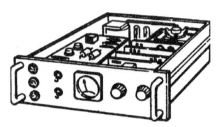

Circuit board assembly
There are three forms of circuit board used in prototype assembly:

● matrix board (1)
● strip board (2)
● copper-clad printed circuit board (3).

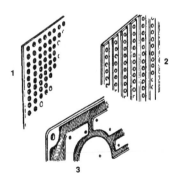

While all three forms are acceptable, company procedure may indicate use of one form (probably printed circuit board) is preferred. To show the three methods, one circuit is shown made up using each form of circuit board.

Component list

Resistors
R1	82k
R2	18k
R3	4k7
R4, R5	470R

Capacitors
C1	1μ, 6V4
C2	2μ, 10V
C3	20μ, 6V4

Semiconductors
VT1	BC108 transistor

Circuit details
Supply V	=	9V
Signal gain	=	x10
Z in	=	15kΩ
Z out	=	5kΩ
Ic	=	1.2mA

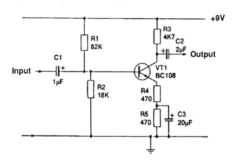

Matrix board

As components are laid out in a similar way to the circuit, it is a simple matter to trace components. Alteration can be easily made.

Method of construction:

● draw a grid on paper. Distance between line intersections should be the same as hole distances on matrix board
● draw components to scale on grid, marking positions of terminal pins. Appearance of drawing should be similar to circuit

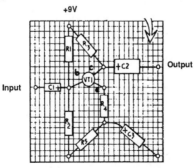

● insert tags into identified holes on matrix board

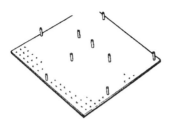

● assemble and solder components to pin tags
● connect leads to appropriate tags.

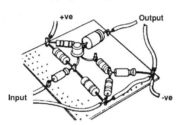

Strip board

Strip board, like matrix board, allows design alterations to be easily incorporated. Layout design, however, does not generally follow circuit appearance, so is slightly harder to follow.

Method of construction:

● letter each horizontal copper strip
● draw onto paper components in approximate positions
● mark necessary breaks in copper tracks

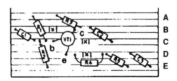

● insert components working to the prepared drawing

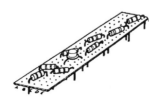

● invert board and solder component leads to tracks. Cut off
excess component leads with side-cutters

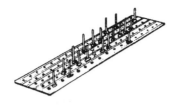

● break copper tracks as required, using strip board cutter
● connect leads to appropriate tracks.

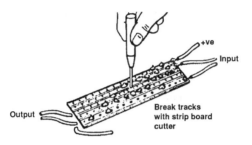

+ve

Input

Output

Break tracks
with strip board
cutter

Printed circuit boards

Often printed circuit boards are constructed as a final stage of
prototype assembly, ie after circuit is proved on matrix or strip
board methods. However, with experience, it is not uncommon
to be able to proceed immediately to printed circuit board form.

A printed circuit board comprises a metallic foil pattern
attached to the surface of an insulator board. Metallic foil makes
up interconnections between components. Pattern of foil is
called track, while component connecting points are called
pads, or lands.

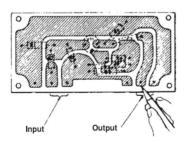

Input   Output

Components are mounted (usually) on the opposite side of
the board to the track. Component leads pass through holes in
the board and are soldered to track pads.

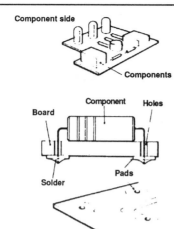

Construction of printed circuit board is covered earlier.

Laying out a front panel
Temporarily layout front panel components on the panel or a drawing of the panel outline.

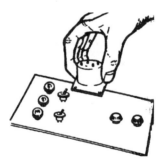

When layout appears satisfactory, discuss work with the circuit designer.

On agreement, sketch front panel layout. It is best to use a 1:1 scale representation.

Carefully mark out fixing holes and slots on the front panel, then make them as required. On prototypes, this marking is usually carried out on the panel rear.

Temporarily attach front panel to chassis. Check for accuracy of fit and remove again.

Mount front panel components as required.

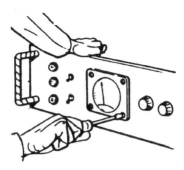

Laying out chassis

Refer to the circuit diagram and temporarily position components on the chassis or a drawing of the chassis outline. Make sure electrical requirements are taken into consideration. For instance, if a transformer is mounted close to an LF choke, windings should be at right angles. Keep components which may get hot well clear of other components.

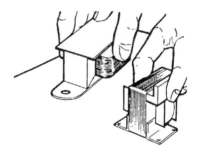

Components likely to produce electromagnetic fields should be positioned well clear of meters on the front panel.

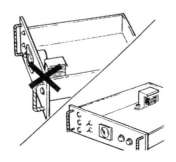

Take care to ensure access to components is unimpeded.

Give thought to chassis layout to ensure no components protruding through the rear of the front panel foul chassis mounted components when assembled. Once again discuss work with the circuit designer.

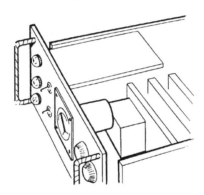

On agreement sketch chassis component layout. It is best to use a 1:1 scale representation.

Mark out and make all fixing holes and slots in the chassis. If specified, chassis should be plated at this stage.

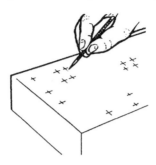

Attach all fixtures such as handbushes, stand-offs and printed circuit connectors.

Secure components on the chassis. Ensure efficient earthing of all metal components. Check signal inputs are routed away from mains leads and stage outputs.

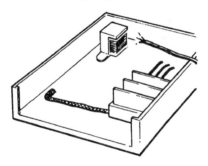

Wiring a chassis
Refer to your sketch of chassis component layout and draw all wiring between components on this layout. Tick each interconnection on the circuit diagram as each lead is drawn on the wiring layout.

Record length, type and colour of each wire on a wiring schedule as it is sketched in position on the wiring diagram.

Use grommets where wires are to pass through the metal chassis.

Following wiring diagram and component layout drawing commence base wiring.

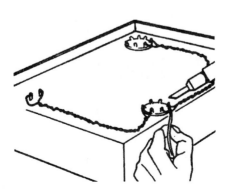

During wiring, check whether some leads would be tidier —
and more economical to insert — as a cableform previously
made as a sub-assembly item.

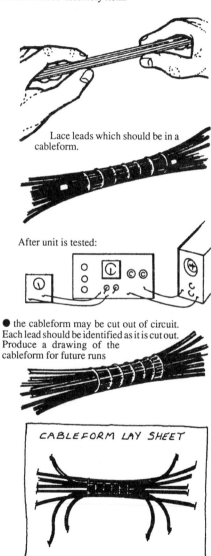

Lace leads which should be in a
cableform.

After unit is tested:

● the cableform may be cut out of circuit.
Each lead should be identified as it is cut out.
Produce a drawing of the
cableform for future runs

CABLEFORM LAY SHEET

● check all tagstrips and tagboards to determine whether they
should be made as sub-assemblies.

When wiring main runs, avoid passing leads over transformers which might radiate interference. Look for leads which can be tidied by combining into a cableform and cleating.

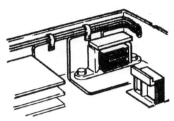

Avoid routing leads across components and boards. Dotted lines show an acceptable — and tidier — arrangement.

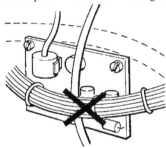

Take particular care not to obscure test points and preset components requiring access at a subsequent stage.

Leads carrying RF or input leads to high gain sensitive circuits should be screened. Form a pigtail for a simple arrangement and solder to an earth tag exposing as little of the centre conductor as possible.

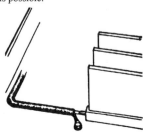

A chassis may contain printed circuit sub-assemblies. These must be assembled and wired before fitting into chassis. Printed circuit boards may have pins for wired connections.

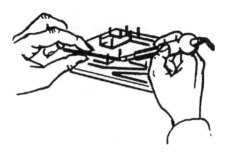

Plug-in printed circuit boards may be used.

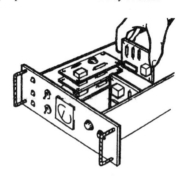

It may be necessary to add further leads to an existing cableform. Temporarily lace extra leads on top of existing lacing — two extra leads are shown.

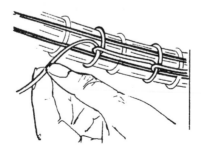

Note changes and record them so cableform drawings can be amended.

Assembly to standards

The General Assembly (GA)

The General Assembly (GA) of a unit with its associated documents completely defines the item being produced. The GA drawing itself shows the finished article and indicates by item numbers various sub-assemblies, components and piece parts. Common fixing parts such as nuts and bolts are shown by item number, adjacent to the item number of one of the parts they secure.

The GA may comprise more than one drawing, in which case drawings have the same number, but are marked sheet 1, sheet 2 and so on. Each sheet shows either a separate stage of assembly or a particular detail of assembly method.

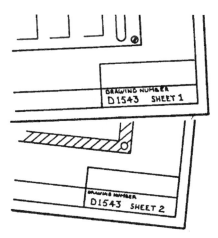

This is the master document and so, in those cases where GA and other documents disagree, the GA is taken as correct. Any seeming error or discrepancy must be reported, using the correct company procedure.

Associated with a GA and often using the same drawing number are other drawings and documents. Typically these include:

- parts list
- drawing list
- circuit diagram
- components list
- wiring diagram
- wiring schedule
- test specification
- component layout.

Read drawings carefully. Do not attempt to use them until they are understood. Read all notes on a drawing:

● note standards and tolerances
● note drawing scale
● obtain and read any specification called up on the drawing.

Issue number
Note drawing issue number. This identifies it from modified versions of the same drawing.

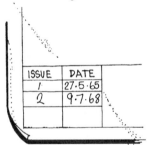

Drawing lists
Drawings for an equipment, and their latest issue numbers, are detailed on the equipment's drawing list. Check with the drawing list to ensure the correct issue number is in use.

SIZE	Drawing No.	ISSUE No	SIZE	Drawing No.
DL	640/1/0R20B	1		
B	SP/8 18662A/3	5		
D	SP/8 18U567	3		

If no drawing list is available, check issue number with the drawing office:

● at the beginning of a new job
● on recommencement of work after a production break.

New issues
When new issues of drawings become available:

● check unmodified parts of new drawings against old drawings
● dispose of old drawings according to company procedure.

Defect reporting
During assembly of prototypes errors or defects in original drawings and documents may be found. These must be reported and drawings amended to suit, so defects will not occur in main production batches.

Documents used to list errors found have many names, depending on company procedure.

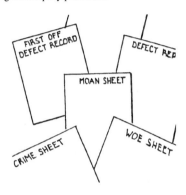

There are many ways of reporting defects but information required is the same:

● nature of defect — a wrong part, an omitted component, wrong value, wrong sequence
● drawing or document in error
● method of correcting error
● shortages or over-supply of items.

DEFECT REPORT			
	DRAWING NUMBER		CORRECTION
1	Part list		Item 6 - Amend qty to 6
2	"		Item 7 Amend length to 7½"
3	"		Add new item: Washers to Spec1234 Qty 2

Use of drawings

For a typical printed circuit board sub-assembly, important
drawings are:

● circuit diagram — shows by symbols the parts of a circuit and
their interconnections. It shows relation of parts in the circuit
and, to a trained eye, operation, but not physical positions of
parts in the sub-assembly

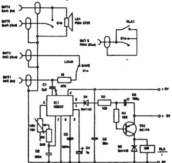

● printed circuit board layout — a drawing showing track of the
printed circuit board

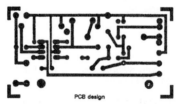

PCB design

● component layout and wiring diagram — indicating positions
of components on the printed circuit board, and wiring
connections to parts not positioned on the board. Sometimes the
component layout and the wiring diagram may be separate

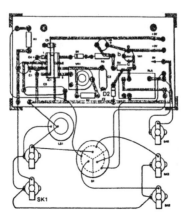

● wiring schedule — where wired interconnections are manifold a supplementary wiring schedule is common, listing interconnections to be made in a strict sequence.

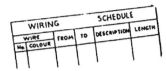

Cabinet fitting and wiring

Cabinets, known also as racks, are made up of a frame into which are fitted a number of self-contained units. Inter-unit wiring joins units, and there are common connections between units and external equipment.

Many sizes of cabinet are available. Usually these are of angle iron or extruded aluminium construction, with standard:

● internal dimensions
● horizontal intervals between tapped holes
● vertical intervals between holes.

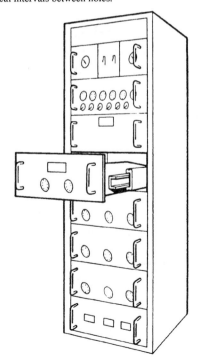

Each supplier provides detailed drawings of drawer mechanism assembly.

Some units are supported on sliding members so units slide forward to a position which allows access for servicing.

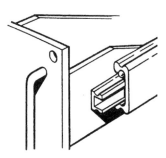

Some sliding mechanisms may be quite elaborate, containing balls which run in slots in top and bottom runners.

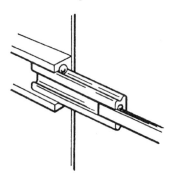

Units are often mounted into racks by bolts. Standard plain washers should be used behind heads of fixing screws of such units.

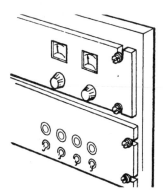

Smaller units may be fitted in cabinet door space.

Inter-unit wiring is carried in cableforms fitted into the rack. Cableform must be fitted so:

- units may be partially withdrawn without straining cables
- cableform can be simply unplugged if units are to be removed
- cables are not trapped by a unit when replaced
- access to unit rear is not impeded

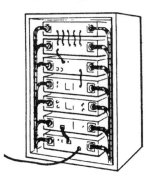

- cable runs are supported close to units, to prevent damage if connectors are removed and connector terminations hang free

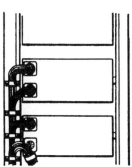

● there are no sharp bends in cableform so no wear against protruding metal is possible.

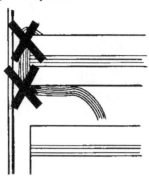

If direct inter-unit connections which are not in a cableform exist, these should be neat and orderly.

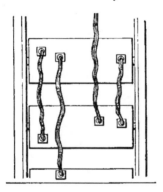

Inter-unit connections must be made with an approved connector method.

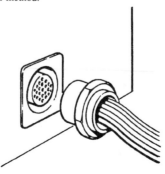

Keep RF cables clear of other circuits. Small co-axial cables may be included in a cableform, however.

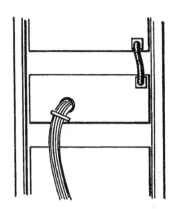

Interconnections between units in doors and racked units should be made by flexible leads arranged in flat arrays. Length of these should be adjusted to prevent strain or trapping and pinching as cabinet doors open and close.

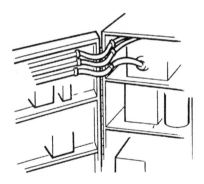

Cabinets must be provided with a main earth terminal, to which all units are earthed.

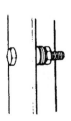

Each unit must be independently connected to the earth terminal by a cable called an earth continuity conductor.

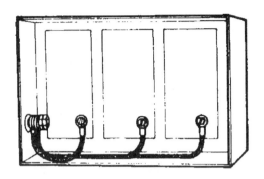

An alternative method of earthing each unit is with a solid copper or aluminium bar behind units.

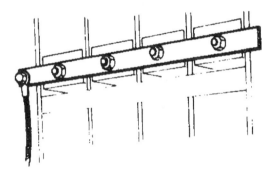

Parts of doors and so on which are not adequately earthed to the main cabinet must be bonded by an earth continuity conductor.

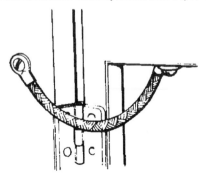

Adhesives

There are many types of adhesive. Which type is used in any application depends on such factors as materials to be adhered, size of mating surfaces, bond strength required, operational atmosphere and so on. Refer to manufacturers' literature to choose particular adhesives.

Safety
● do not allow adhesives to come into contact with skin
● wash off immediately any resin, hardener, or other forms of adhesive coming into contact with skin
● do not inhale fumes

● many adhesives are highly flammable. Keep away from flames, electric fires, cigarettes, sparking electrical equipment and so on
● remove any accidental contamination immediately, using a suitable solvent
● use a suitable barrier cream on hands before using adhesive
● use rubber gloves during prolonged use of adhesive.

General-purpose adhesive

These are used to secure light materials where there are no special conditions to be considered. It is usually supplied in single containers, and is applied between items which are to be secured.

Epoxy adhesive

Provides a strong clear bond between most hard materials, which is not affected by moderate heat and moisture. It is supplied in two parts, a resin and a hardener, which must not be mixed until required for use. To mix and use epoxy adhesive:

● mix together appropriate quantities of resin on a clean, dry surface

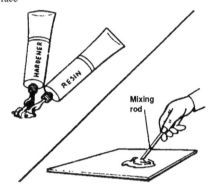

● apply mixture to joint with a spatula
● clamp joint firmly in position and remove excess adhesive
● allow adhesive to cure (see manufacturer's instructions) before removing clamp.

Impact (contact) adhesive

Used to bond large areas of flat or preformed materials. To apply impact adhesive:

● clean surfaces to be joined using appropriate solvent where specified
● apply a thin even layer of adhesive to each surface

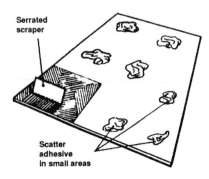

● allow adhesive solvent to evaporate so both surfaces become tacky
● press surfaces together in required position.

Note there is no possibility of movement once surfaces are mated so it is essential parts are brought together in required position. Some varieties of impact adhesive, however, do allow minor movement after mating.

Epoxy resins
Resin is typically a syrupy liquid which, when mixed with a hardener (catalyst, curing agent), is rapidly transformed into a hard transparent solid.

When selecting resin and hardener, it should be borne in mind that resins are available which will cure at temperatures varying from room temperature to something in excess of 140°C. Proportions of resin and hardener are critical, so where large amounts of adhesive is to be mixed, weighing is the best method.

Materials to be joined
Epoxy resins form extremely strong and durable bonds with metals, glass, rigid plastics, rubber, wood and many other materials.

Surface preparation
Bond strength depends on thoroughness with which surfaces to be joined are cleaned. All dirt and grease must be removed. Refer to manufacturers' literature for details of surface preparation.

Hand tools

Spanners
Double-ended box spanner
Made from tube steel, sometimes alloy steel such as chrome molybdenum. A box spanner is used where position of a nut or bolt is such that is cannot be reached by an open-ended or ring spanner. It is identified by size in the same way as other spanners.

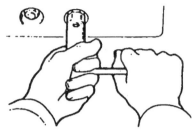

Hexagon socket screw wrenches

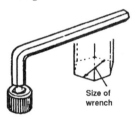

Used for hexagon socket screw, they are identified by size across flats.

Size of wrench

Spanner defects

Important
A defective spanner must never be used. Its condition must always be checked before it is used.

Spanners and common defects are:

● ring spanners — worn and rounded-off internal serrations. Spanners slip under pressure

● box spanners — worn and rounded-off hexagon shaped ends, and split corners of hexagons. Spanners slip under pressure

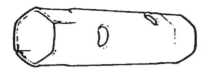

● socket sets — worn and rounded-off serrations in sockets, causing spanner slip, also worn squares and locking devices on wrench arms

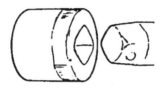

● torque wrenches — faulty torque unit giving incorrect torque loading
● hexagon socket screw wrenches — worn and rounded-off ends. Wrenches slip under pressure

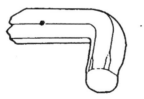

● adjustable wrenches — sprung jaws, worn screw jack. Wrenches slip under pressure.

Worn jack

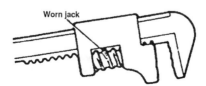

Correct use of spanners

Correct sizes of spanner must always be used. Attempts to improvise invariably cause damage and may be dangerous.

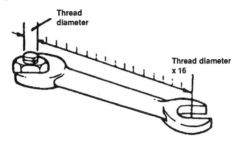

A spanner must not be lengthened by adding other tools to increase leverage.

Screwdrivers

Electrician's screwdrivers

Obtained in sizes from 100 mm up to 250 mm blade length. Handles are insulated against high voltage, so these should be used whenever work is undertaken on electrical equipment.

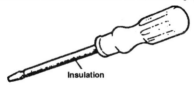

Safety

Never use an uninsulated screwdriver when working on electrical circuits.

Pliers

Pliers fall into two categories: those for gripping, used to hold small components which would otherwise be difficult to control, and those for cutting, with sharpened jaws. Pliers are classified by their usage:

● end cutting — for cutting wire

● side cutting — for cutting wire and insulation

● sniped-nose — for holding small items

● rounded-nose — for looping wire ends

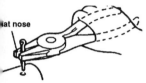

lat nose

● flat-nose — used solely for gripping and holding purposes

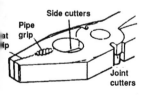

Side cutters
Pipe grip
at
ip
Joint cutters

● combination — versatile tools with a large number of applications. They incorporate side cutters, joint cutters and pipe grip, in addition to general-purpose flat grip

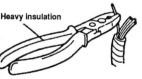

Heavy insulation

● heavy duty electrical pliers — similar to combination pliers with addition of insulated handles. They should always be used when working with electrical components.

Inspection of pliers
Pliers are subject to following defects:

● worn fulcrum pin — jaws should open and close smoothly without any sloppiness or sideplay between jaws

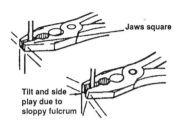

● worn serrations and sprung jaws — this reduces gripping power and may result in damage to the workpiece

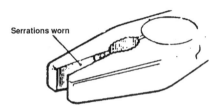

● damaged side cutters — this gives difficulty when cutting wire or pins to a suitable length.

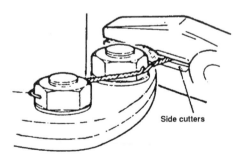

Important
Condition of pliers should always be be checked before they are used. Defective pliers must never be used.

Specialised pliers

There are various other types of pliers which have been developed for special applications. Some of these include:

● pliers for external circlips

● pliers for internal circlips

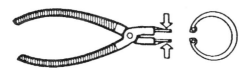

● two-way circlip pliers — by squeezing appropriate handles the steel points open or close as required

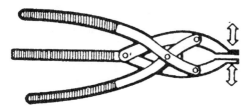

● a special circlip tool, for larger clips in confined spaces, such as bearing housings on larger machines. Screw action holds the circlip without continuous hand pressure

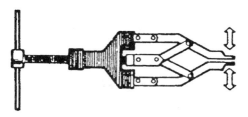

● sleeve fitting pliers — used for fitting sleeves to terminals and joints, or to fit identification bands to cables.

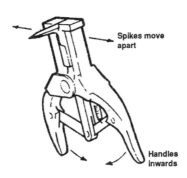

Spikes move apart

Handles inwards

Wire insulation strippers
Mechanical strippers
A number of types of mechanical strippers exist:

● simple strippers — two plate arrangement with central fulcrum. Notches in the plates cut through insulation

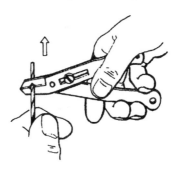

● precision strippers — with V-shaped notches in accurately adjustable jaws

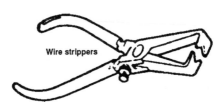

Wire strippers

● precision strippers — with mechanical action which cuts and strips insulation in one action.

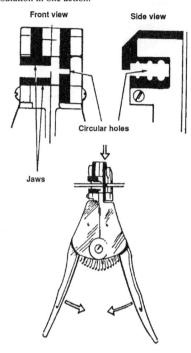

Front view

Side view

Circular holes

Jaws

Thermal strippers
This tool allows insulation to be removed by melting through the insulation prior to pulling the tool away from the wire.

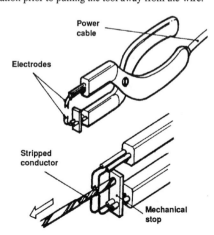

Power cable

Electrodes

Stripped conductor

Mechanical stop

Marker tool
Used to fit marker sleeves to wires.

Pull marker and wire

Bullet tool
Used to fit marker sleeves to wires.

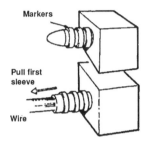

Markers

Pull first
sleeve

Wire

Tweezers
Tweezers are used for manipulation and holding small components or wire leads when soldering.

Pop-riveting tools
These provide a convenient way of riveting light gauge materials (up to about 3 mm) where air or watertight joints are not required. Access to only one side of the joint is required. To use:

● select a suitable-sized rivet, and drill material to be riveted
● locate rivet in the hole, hold materials in required position and squeeze handles of tool

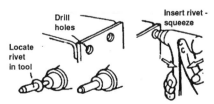

Drill
holes

Insert rivet -
squeeze

Locate
rivet
in tool

● remove head of broken pin if there is a possibility it could cause a short circuit.

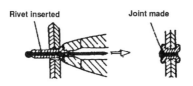

Rivet inserted   Joint made

Portable electric tools

Safety
Users of portable electric tools should be familiar with correct operating procedures.

Users should be familiar with treatment of electric shock.

Do not use a portable electric tool near locations where petrol, cellulose or other unsafe materials or atmospheres are in usage or storage. Sparks from a tool's commutator could cause an explosion or start a fire.

Check a tool's condition before use. A faulty electric tool must never be used. If there is slightest doubt about an electric tool's internal condition it must not be used.

Report immediately any defect noticed in a tool, cable or plug.

A power tool's voltage must always be checked before use. Plugs must be securely fixed in the nearest power point, and the lead should be routed to the job in the safest way.

Metallic cased tools
Portable electric tools with metallic outer casings must have their casings earthed. Socket outlets and plugs must be of a 3-pin type: live, neutral and earth. Power supply cables must be

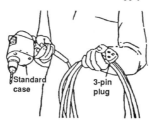

Standard   3-pin
case   plug

3-cored and of a size specified for their function. Earth continuity must be checked frequently.

Safety
Never connect such tools to a 2-wire supply. It is dangerous practice and is contrary to factory regulations

Double-insulated tools

Fully-insulated or double-insulated portable tools are available. These tools are designed and constructed so no metal part can become live accidentally. Earthing of the tool is therefore unnecessary and a 2-cored power supply cable is used.

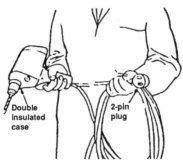

Transformer supply

When portable power tools are used on site they should have a working voltage of 110V, supplied by a centre-tapped transformer. Centre tap must be connected to earth so supply voltage is only 55V above earth potential.

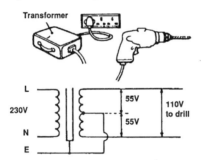

Inspecting power tools before use

Check for defects:

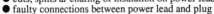

● cuts, splits ar chafing of insulation on power lead
● faulty connections between power lead and plug

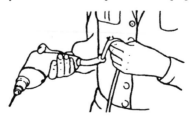

● damaged or blocked screens at air inlets, resulting in a damaged or burned out motor — blocked air outlets have the same effect

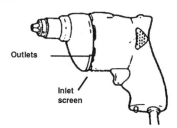

Outlets

Inlet screen

● damaged casing — cracks or looseness. A tool may have been dropped with resulting internal damage.

Note all power tools other than double-insulated types must have 3-core power leads.

Portable electric drills
These are hand-held drilling machines in which a drill spindle is driven by an electric motor driving through a reduction gear train.

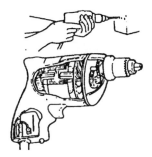

When using a portable electric drill, it is important to ensure:

● drill bit is correctly located in the chuck, and is tightly secured and running true before it is applied to the work

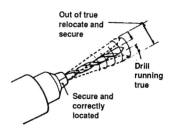

Out of true relocate and secure

Drill running true

Secure and correctly located

● pressure applied is never so great that speed of motor rotation
is appreciably reduced. At point of breakthrough, pressure
should be relieved to ensure a smooth clean breakthrough.

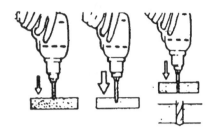

Measuring instruments

Electrical measuring instruments are usually based around
delicate rotary movements. A number of precautions should be
followed at all times:

● handle instruments and meters carefully. They are delicate
and expensive
● place instruments in a safe position where they cannot fall
● avoid trailing test leads over bench edges or when carrying
● carry instruments by handles or straps provided
● avoid applying pressure to the glass
● do not tap the glass
● report any damage. Damaged instruments give inaccurate
measurements

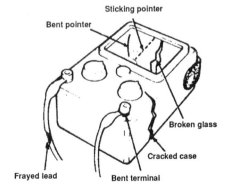

● make sure you know how to use an instrument correctly. Read
the manufacturer's instruction book

● when using a meter, position it as specified by the manufacturer — this is the position in which it has been calibrated
● keep meters away from magnets and ferrous materials as these affect meter readings
● before using a meter, check the pointer reads zero. If necessary, adjust to zero using the pointer adjusting screw.

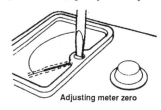

Adjusting meter zero

Ohmmeters and multimeters
Ohmmeter and multimeters contain cells or batteries. If it is not possible to calibrate a meter's resistance range, cell or battery deterioration is the most likely cause. Observe cell or battery polarity when refitting.

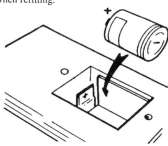

Before connecting a multimeter to a circuit check:

● function switch is set correctly
● meter is switched to highest range for initial reading. Switch to a lower range only when you are sure the meter will not be overloaded.

Never switch a meter to current or resistance ranges when connected to a voltage. After use, switch a meter to its highest AC range.

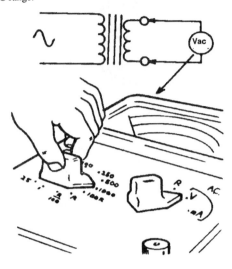

Characteristics of measuring instruments

There are two types of basic movement in general use:

● moving iron — used to measure direct and alternating current. Such a meter has an uneven scale

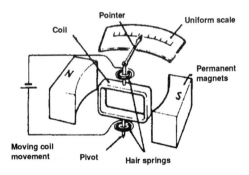

● moving coil — measures direct currents up to about 75 mA and direct voltages up to about 75 mV, although modern multirange instruments often have movements which operate at currents as small as 100 µA. Moving coil meters have uniform scales.

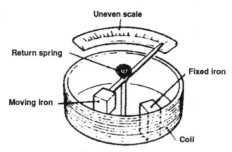

Moving iron movement (repulsion type)

To measure currents greater than this, moving coil meters use low resistance shunts across the movement. A shunt bypasses most of the current, allowing only a small, known, proportion of the total to pass through the coil of the movement.

Electronic test equipment

While a visual examination of an electronic assembly is usually sufficient to reveal faults such as overheated components or broken connections, more obscure faults must usually be located using equipment specially made for the purpose. Such test equipment is varied, ranging from simple meters through to highly complex and specific automatic test equipment.

Automatic test equipment, however, being so specific, is designed and built to test one assembly type — that is, when a new assembly is produced, new automatic test equipment is normally required.

Moving coil movement with rectifier

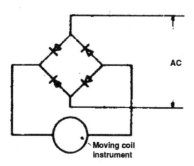

Between simple meters and complex automatic test equipment is a range of electronic test equipment which is general-purpose, so can be used to test many different assembly types. These fall into the categories of:

● cathode ray oscilloscope
● signal generator
● digital multimeter
● power supply.

Cathode ray oscilloscope

A cathode ray tube (CRT) is the heart of the cathode ray oscilloscope. Electrons emitted from a cathode within the tube are focused into a thin stream under the effects of a voltage applied at the first anode, then accelerated to a fluorescent screen. They are usually returned to the cathode via a graphite coating on the tube inside.

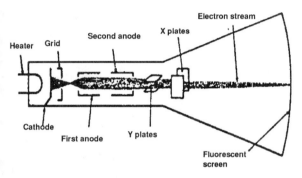

A luminous spot, visible from outside the tube, is formed when electrons strike the screen. Spot intensity is controlled by voltage applied to a grid.

The electron stream is deflected by a magnetic field, produced by either coils around the tube neck, or (more usually) internal electrostatic plates. Two pairs of plates, at right angles, are positioned around the electron stream. Potential difference across each pair governs deflection: higher the potential difference, greater the deflection. One pair of plates (Y plates) is used to deflect the stream vertically, other pair (X plates) is used to deflect it horizontally.

For some tests a timebase circuit controls potential difference across X plates, with a repetitive ramping voltage, which moves spot across screen in a uniform left-to-right movement before quickly flying it back to restart. At most spot movement speeds effect is of a continuous line across the screen.

An input signal, after amplification, usually forms the potential applied to Y plates. Thus input signal amplitude affects electron stream vertical position on the screen.

Combined effect of a timebase signal on X plates and an input signal on Y plates can be a waveform displayed on the oscilloscope screen. Some oscilloscopes allow two separate input signals to share and control the electron stream, effectively splitting it into two and forming two on-screen traces: one for each input. Such dual-trace oscilloscopes are useful for comparison of two signals.

Safety

Oscilloscopes have high internal voltages, some in the kilovolt range, which can discharge through an air gap and give an electric shock without live parts being touched.

These high voltages may be retained by an oscillocope's internal capacitors long after the oscilloscope has been isolated from mains. Therefore, do not remove an oscilloscope's case unless adequate safety precautions have been undertaken.

An oscilloscope case must normally be earthed, either via the earth lead in the mains cable, or by a separate connection from the earth terminal on the case to true earth. If the case is not properly earthed, there is a risk of:

● electric shock, if the case becomes live
● spurious readings due to ineffective screening from interference.

When using an oscilloscope to test live equipment:

● check equipment under test to see if its chassis is live to earth

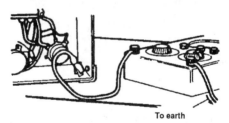

To earth

● when connecting test leads to live circuits, use only insulated clips or prods and keep hands well away from live components.

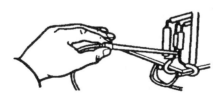

The oscilloscope is a delicate instrument. When using it:

● avoid damage through mechanical shock when man-handling
● keep brightness to a minimum effective setting when in use; when not in use but switched on, turn brightness to zero. This will help prevent burning of the screen, which occurs particu-

larly if the trace is stationary. An oscilloscope's on/off switch
is often incorporated in the brightness control.

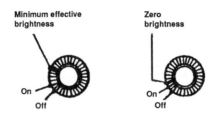

Setting up before use
Before switching on an oscilloscope:

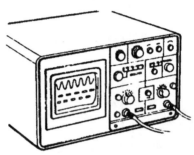

● set vertical amplifier coarse gain to minimum, to avoid
overload damage
● set focus, X shift and Y shift to mid-positions
● set brightness and X gain to minimum
● set timebase to a suitable position
● set trigger selection to internal trigger
● set stability to auto/normal
● select AC position.

When switching on:

● switch on and allow to warm up
● advance brightness until a trace appears
● centralise trace with X and Y shift controls
● adjust focus for a clear sharp trace.

Displaying a waveform
One use of an oscilloscope is to display a waveform from a
circuit under test. Procedure is:

● switch to AC input
● connect input leads. Unless the oscilloscope has a differential
input one of the leads may be connected only to an earthed point
in the tested circuit

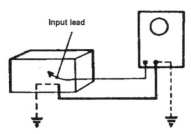

● increase Y gain to obtain a convenient waveform size

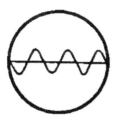

● adjust timebase until the waveform is clearly displayed.

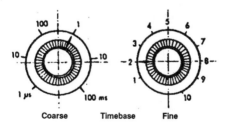

Coarse   Timebase   Fine

Measuring frequency
Once a waveform is displayed, its frequency can be measured because timebase is calibrated (in time per centimetre eg, microseconds/cm), to give time taken by the trace to move horizontally over one centimetre on the screen. To measure wave frequency:

● adjust timebase so a whole number of cycles occupies a whole number of centimetres on the screen

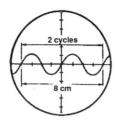

● calculate frequency from the expression:

$$\text{frequency} = \frac{\text{number of cycles}}{\text{time/cm} \times \text{number of cms}}$$

In the screen illustration:

$$\text{frequency} = \frac{2}{2 \times 10^{-3} \times 8} = 125 \text{ Hz.}$$

Measuring voltage
Some oscilloscopes have directly coupled amplifiers which enable them to measure both direct and alternating voltages.

To measure alternating voltage:

● switch to AC

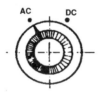

● adjust Y shift to set trace on zero line of screen's vertical scale

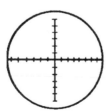

● connect test leads to voltage source
● move fine Y gain control to calibrate position

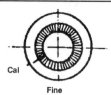

Cal

Fine

● adjust coarse Y gain control to bring waveform height to a convenient scale position, say, about three-quarters of maximum scale reading

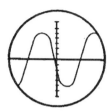

● adjust X shift to bring the peak of a wave on the vertical scale line. Note reading

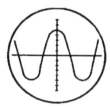

● repeat this for a wave trough

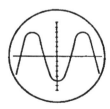

● calculate peak-to-peak voltage from Y gain setting and wave height in centimetres, where peak-to-peak voltage is number of centimetres multiplied by volts/cm

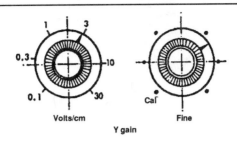

Y gain

● for a sinewave, voltage is calculated from the expression:

$$\text{RMS voltage} = 0.707 \times \frac{\text{peak-to-peak volts}}{2}$$

To measure direct voltage
A direct voltage, or a DC component of an alternating voltage
can be measured by:

● switching oscilloscope to DC, and allowing an adequate
warm-up time to prevent drift between measurements

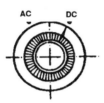

● shorting test leads together, adjust Y shift to position trace on
zero line of vertical scale

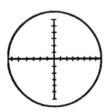

● measure DC voltage. DC component of an AC waveform is
measured by reading to the centre line of the waveform.

Signal generators
These produce a sine, square or triangular waveform output at
a variable frequency and amplitude. Value of frequency and
amplitude is indicated on a digital display, or on calibrated
adjustment dials.

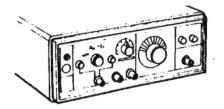

Signal generators are frequently used to provide input to, say, an amplifier whose amount of distortion is being measured, or whose frequency range is being determined.

Signal generators are categorised as general-purpose or standard signal types according to:

● accuracy of frequency and amplitude
● stability of frequency and amplitude
● distortion present in output.

Check in manufacturers' instruction books to see if instrument performance is suitable for a particular measurement to be made. Use a general-purpose instrument where accuracy and stability are not important. Use a standard signal generator where accuracy and stability are important.

Radio frequency signal generators
A radio frequency (RF) signal generator provides:

● a continuous wave RF signal
● a modulated continuous wave RF signal.

Instruments are available which cover frequencies of 100 kHz and upwards. Use an RF signal generator to test, service and measure performance of equipment operating in RF.

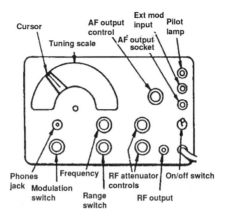

Audio frequency signal generators

An audio frequency (AF) signal generator provides an AF signal of sinewave output, squarewave output, or both. Sometimes triangular wave output is included.

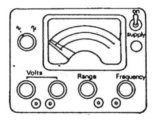

A typical AF signal generator covers the frequency range 5 Hz to 100 kHz at an output voltage adjustable from 0 to 30V. Use an AF signal generator to test, service and measure performance of equipment operating in the AF band.

Digital multimeters

These instruments have a high input impedance and are therefore ideal for voltage measurements where current available is so low an analogue meter of moving coil type would give false readings.

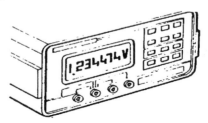

Index